ENGINEERING MATERIALS

An Introduction to their Properties and Applications

Other titles of interest

ASHBY	Materials Selection in Mechanical Design
ASHBY & JONES	Engineering Materials 2
GIBSON & ASHBY	Cellular Solids
HULL & BACON	Introduction to Dislocations, 3rd Edition
JONO & INOUE	Mechanical Behaviour of Materials – VI (ICM 6)
KETTUNEN *et al.*	Strength of Materials and Alloys (ICSMA 8)
SALAMA *et al.*	Advances in Fracture Research (ICF 7)
TAYA & ARSENAULT	Metal Matrix Composites: Thoromechanical Behaviour
YAN *et al.*	Mechanical Behaviour of Materials (ICM 5)

ENGINEERING MATERIALS

An Introduction to their Properties and Applications

by

MICHAEL F. ASHBY

and

DAVID R. H. JONES

Cambridge University, England

Butterworth-Heinemann Ltd
Linacre House, Jordan Hill, Oxford OX2 8DP

A member of the Reed Elsevier plc group

OXFORD LONDON BOSTON
MUNICH NEW DELHI SINGAPORE SYDNEY
TOKYO TORONTO WELLINGTON

First published by Pergamon Press Ltd 1980
Reprinted with minor corrections 1981
Reprinted 1985, 1986, 1987, 1988, 1989, 1991, 1993, 1995

British Library Cataloguing in Publication Data
Ashby, M. F.
Engineering materials 1 – (International series on materials science and technology)
1. Materials
I. Title II. Jones, D. R. H.
620.1'1 TA403 80–40623

ISBN 0 7506 2766 2

Printed and bound in Great Britain by BPC Wheatons, Exeter

CONTENTS

GENERAL INTRODUCTION

To the student

Innovation in Engineering often means the clever use of a new material—new to a particular application, but not necessarily (although sometimes) new in the sense of "recently developed". Plastic paper clips and ceramic turbine-blades both represent attempts to do better with polymers and ceramics what had previously been done well with metals. And engineering disasters are frequently caused by the misuse of materials. When the plastic tea-spoon buckles as you stir your tea, and when a fleet of aircraft is grounded because cracks have appeared in the tailplane, it is because the engineer who designed them used the wrong materials or did not understand the properties of those he did use. So it is vital that the professional engineer should know how to select materials which best fit the demands of his design—economic and aesthetic demands, as well as demands of strength and durability. He must understand the properties of his materials, and their limitations.

This book gives a broad introduction to these properties and limitations. It cannot make you a materials expert, but it can teach you how to make a sensible choice of material, how to avoid the mistakes that have led to embarrassment or tragedy in the past, and where to turn for further, more detailed, help.

You will notice from the Contents list that the chapters are arranged in *groups*, each group describing a particular class of properties: the Elastic Modulus; the Fracture Toughness; Resistance to Corrosion; and so forth. Each such group of chapters starts by *defining the property*, describing how it is *measured*, and giving a table of *data* that we use to solve problems involving the selection and use of materials. We then move on to the *basic science* that underlies each property, and show how we can use this fundamental knowledge to design materials with better properties. Each group ends with a chapter of *Case Studies* in which the basic understanding and the data for each property are applied to practical engineering problems involving materials. Each chapter has a list of books for *further reading*, ranked so that the more elementary come first.

At the end of the book you will find sets of examples; each example is meant to consolidate or develop a particular point covered in the text. Try to do the examples that derive from a particular chapter while this is still fresh in your mind. In this way you will gain confidence that you are on top of the subject.

No engineer attempts to learn or remember tables or lists of data for material properties. But you *should* try to remember the broad orders-of-magnitude of these quantities. All grocers know that "a kg of apples is about 10 apples"—they still weigh them, but their knowledge prevents them making silly mistakes which might cost them money. In the same way, an engineer should know that "most elastic moduli lie between 1 and $10^3\,\mathrm{GN\,m^{-2}}$; and are around $10^2\,\mathrm{GN\,m^{-2}}$ for metals"—in any real design you need an accurate value, which you can get from suppliers' specifications; but an order-of-magnitude knowledge prevents you getting the units wrong, or making other silly, and possibly expensive, mistakes. To help you in this, we have added at the end of the book a list of the important definitions and formulae that you should know, or should be able to derive, and a summary of the orders-of-magnitude of materials properties.

To the lecturer

This book is a course in Engineering Materials for engineering students with no previous background in the subject. It is designed to link up with the teaching of Design, Mechanics and Structures, and to meet the needs of engineering students in the 1980s for a first materials course, emphasising applications.

The text is deliberately concise. Each chapter is designed to cover the content of one 50-minute lecture, twenty-seven in all, and allows time for demonstrations and illustrative slides. A list of the slides, and a description of the demonstrations that we have found appropriate to each lecture, are given in Appendix 2. The text contains sets of worked case studies (Chapters 7, 12, 16, 20, 22, 24, 26 and 27) which apply the material of the preceding block of lectures. There are examples for the student at the end of the book; worked solutions are available separately from the publisher.

We have made every effort to keep the mathematical analysis as simple as possible whilst still retaining the essential physical understanding, and still arriving at results which, although approximate, are useful. But we have avoided mere description: most of the case studies and examples involve analysis, and the use of data, to arrive at numerical solutions to real or postulated problems. This level of analysis, and these data, are of the type that would be used in a preliminary study for the selection of a material or the analysis of a design (or design-failure). It is worth emphasising to students that the next step would be a detailed analysis, using *more precise mechanics* (from the texts given as "further reading") and *data from the supplier of the material or from in-house testing*. Materials data are notoriously variable. Approximate tabulations like those given here, though useful, should never be used for final designs.

CHAPTER 1

ENGINEERING MATERIALS AND THEIR PROPERTIES

Introduction

In designing a structure or device, an engineer has a vast range of materials at his disposal. How does he go about selecting the material, or combination of materials, which best suits his purpose? Mistakes can cause disasters. During the last war, for example, one class of welded merchant ship suffered heavy losses, not by enemy attack, but by breaking in half at sea. Failure occurred by a fracture running along the welds and right around the ship because the weld material had a low *fracture toughness*. This—a bulk mechanical property of the material—is listed in Table 1.1 along with the other common classes of property that the designer must consider when choosing his materials. Many of these classes of property will be unfamiliar to you—we shall only be referring to them by way of example in this chapter—and they form the basis of this course on materials. During our course, we shall also encounter a number of classes of materials (shown in Table 1.2) from which we can make our artefacts.

In this chapter we illustrate, through a variety of examples, how the designer selects his materials so that they provide him with the properties he requires. Our first example is the selection of materials for a

Plastic-handled screwdriver

A typical screwdriver has a shaft and blade made of a high-carbon steel, a metal. Steel is chosen because its *modulus* is high. The modulus measures the resistance of the material to elastic deflection or bending. If you made the shaft out of a polymer like polyethylene instead, it would twist far too much. This property (the modulus) is one criterion in the selection of a material for this application. But it is not the only one. The shaft must have a high *yield strength*. If it does not, it will bend plastically or permanently if you twist it hard (bad screwdrivers do). And the blade must have a high *hardness*, otherwise it will be indented by the material of the head of the screw, and thereby damaged. Finally, the material of the shaft and blade must not only resist bending and twisting, it must also resist fracture—glass, for instance, has a high modulus, yield strength and hardness, but it would obviously be a bad material for this application, because it is so brittle. More precisely, it has a very low *fracture toughness*. That of the steel is high, meaning that it gives, or bends, before it breaks.

TABLE 1.1
CLASSES OF PROPERTY

Property	Class
Price and availability	Economic Properties
Density Modulus and damping Yield strength, tensile strength, hardness Fracture toughness Fatigue strength, thermal fatigue resistance Creep strength	Bulk mechanical Properties
Thermal properties Optical properties Magnetic properties Electrical properties	Bulk non-mechanical properties
Oxidation and corrosion Friction, abrasion and wear	Surface properties
Ease of manufacture, Fabrication, joining, finishing	Production properties
Appearance, texture, feel	Aesthetic properties

The handle of the screwdriver is made of a polymer or plastic, in this instance polymethylmethacrylate, otherwise known as PMMA or perspex. The handle has a much larger section than the shaft, so its twisting, and thus its modulus, is less important. You could not make it satisfactorily out of a rubber (another polymer) because its modulus is much too low, but most other polymers could be used. Traditionally, of course, tool handles were made of another, rather complicated, polymer—wood—and, if you measure importance by the tonnage consumed per year, wood is still by far the most important polymer available to the engineer. Wood has been replaced by PMMA because PMMA becomes soft when hot and can be moulded quickly and easily to its final shape. Its *ease of fabrication* for this application is high. It is also chosen for aesthetic reasons: its *appearance*, and feel or *texture*, are right; and its density is low, so that the screwdriver is not unnecessarily tiring to hold. Finally, PMMA is cheap, and this allows the product to be made at a reasonable *price*.

Our second example takes us from low technology to the advanced materials design involved in the Rolls-Royce RB211 turbofan aero-engine as used to power the "Jumbo-Jets". Air is compressed into the engine by the turbofan, which also provides aerodynamic thrust around the outside of the casing. The air is further compressed by the compressor blades, and is then mixed with fuel and burnt in the combustion chamber. The expanding gases drive the engine blades, which provide power to the turbofan and the compressor blades, and finally pass out of the rear of the engine, contributing to the thrust.

The *turbofan blades* are made from a titanium alloy, a metal. This has a sufficiently good modulus, yield strength, and fracture toughness. But the metal must also resist *fatigue* (due to rapidly fluctuating loads), *surface wear* (from striking water droplets at high speed) and *corrosion* (important when taking off over the sea). Finally, *density* is extremely important for obvious reasons; titanium alloys are the lightest available. In an effort to reduce weight even further, composite blades of CFRP, with density less than one-half of that of titanium, have been tried. But CFRP is simply not tough

TABLE 1.2
CLASSES OF MATERIALS

Classes of materials
Metals and alloys
Iron and steels
Aluminium and its alloys
Copper and its alloys
Nickel and its alloys
Titanium and its alloys
Polymers
Polyethylene (PE)
Polymethylmethacrylate (PMMA, Perspex)
Nylon
Polystyrene (PS)
Polyurethane (PU)
Polyvinylchloride (PVC)
Rubbers
*Ceramics and glasses**
Alumina (Al_2O_3, emery, sapphire)
Magnesia (MgO)
Silica (SiO_2) glasses and silicates
Silicon carbide (SiC)
Silicon nitride (Si_3N_4)
Cement and concrete
Composites
Wood
Fibreglass (GFRP)
Carbon-fibre reinforced polymers (CFRP)
Filled polymers
Cermets

* Ceramics are crystalline, inorganic, non-metals. Glasses are non-crystalline (or *amorphous*) solids. Most engineering glasses are non-metals, but a range of *metallic glasses* with useful properties is now available.

enough for turbofan blades—"bird strikes" would have been capable of instantly demolishing a CFRP turbofan.

Turning to the *engine blades*, even more material requirements must be satisfied. For economy the fuel must be burnt at as high a temperature as possible. The first row of engine blades (the "HPI" blades) nowadays runs at metal temperatures of about 950°C. This adds resistance to *creep* and *oxidation* to the above requirements. Nickel-based alloys of enormously complicated make-up are used for this exceedingly stringent application and represent one pinnacle of advanced materials technology.

As an example which brings in somewhat different requirements, the *sparking plug* of an internal combustion engine is interesting. The *spark electrodes* have to be resistant to *thermal fatigue* (from rapidly fluctuating temperatures), *wear* (from spark erosion) and *oxidation and corrosion* from hot upper-cylinder gases containing nasty compounds of sulphur, lead (from anti-knock additives) and so on. Tungsten alloys are used for the electrodes because they have the desired properties.

The *insulation* around the central electrode is an interesting example of a non-metallic material—in this case, alumina, a ceramic. This is used because of its electrical

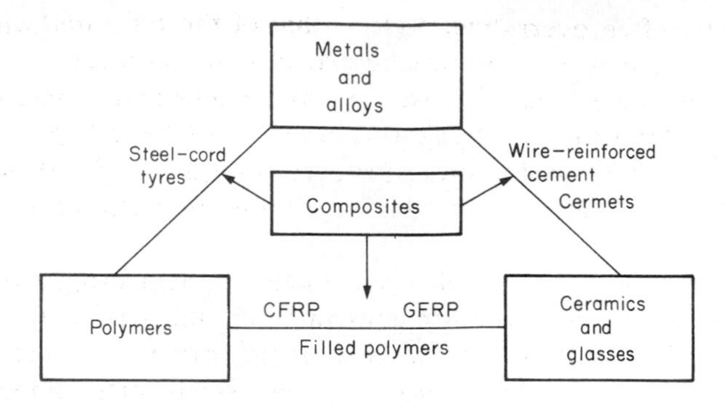

Fig. 1.1. The classes of engineering materials from which articles are made.

insulating properties and because it also has good thermal fatigue resistance and resistance to corrosion and oxidation (it is an oxide already!).

The use of non-metallic materials has become most widespread in the consumer industry. Our next example, a sailing cruiser, shows just how extensively polymers and man-made composites have replaced the "traditional" materials of steel, wood and cotton. A typical cruiser has a *hull* made from GFRP which is easy to manufacture as a complete moulding which has good *appearance* and which, unlike steel or wood, does not rust or become eaten away by Terido worm. The *mast* is made from aluminium alloy, which is much lighter for a given strength than wood; advanced masts are now being made by reinforcing the alloy with boron "whiskers" (a man-made composite). The sails, formerly of the natural material cotton, are now made from the polymer Terylene (with nylon for spinnakers) and, in the running rigging, cotton ropes have been replaced by polymers also. Finally, polymers like PVC are extensively used for things like fenders, anoraks, bouyancy bags and boat covers.

Three man-made composite materials have appeared in the items we have considered so far: glass-fibre reinforced polymers (GFRP); the much more expensive carbon-fibre reinforced polymers (CFRP); and the still *more* expensive boron-fibre reinforced alloys (BFRP). The range of composites is a large and growing one (Fig. 1.1); during the next decade composites will, increasingly, compete with steel and aluminium in many traditional uses of these metals.

So far we have introduced in reasonable detail the mechanical and physical properties of the engineering materials available to us, but we have yet to discuss a very

TABLE 1.3

		Price per tonne
Basic construction	Wood, concrete, structural steel	UK£30–250, US\$60–550
Medium and light engineering	Metals, alloys and polymers for aircraft, automobiles, appliances, etc.	UK£250–2500, US\$550–5500
Special materials	Turbine-blade alloys, advanced composites (CFRP, BFRP), etc.	UK£2500–90,000, US\$5500–200,000
Precious metals, etc.	Sapphire bearings, silver contacts, gold microcircuits	UK£90,000–1 m, US\$200,000–2.2 m
Industrial diamond	Cutting and polishing tools	UK£400 m, US\$900 m

important, and very often overriding, factor—that of the *price and availability* of our materials. Table 1.3 shows a rough breakdown of material prices.

Materials for large-scale structural use—wood, cement and concrete, and structural steel—cost getween UK£30 and UK£250 (US$60 and US$550) per tonne. There are many materials which have all the other properties required of a structural material—nickel, or titanium, for example—but their use in this application is eliminated by their price.

The value that is added during light- and medium-engineering work is larger, and this usually means that the economic constraint on the choice of materials is less severe—a far greater proportion of the cost of the structure is that associated with labour or with production and fabrication. Stainless steels, most aluminium alloys and most polymers cost between UK£250 and UK£2500 (US$550 and US$5500) per tonne. It is in this sector of the market that the competition between materials is most intense, and the greatest scope for imaginative design exists. Here polymers and composites compete directly with metals, and new structural ceramics (silicon carbide and silicon nitride) may compete with both in certain applications.

Next there are the materials developed for high-performance applications or special uses, such as nickel alloys (for turbine blades), tungsten (for sparking-plug electrodes) and special composite materials such as CFRP. The price of these materials ranges between UK£2500 and UK£90,000 (US$5500 and US$200,000) per tonne. This the régime of high materials technology, actively under research, and in which major new advances are continuing to be made. Here, too, there is intense competition from new materials.

Finally, there are the so-called precious metals and gemstones, widely used in engineering: gold for microcircuits, platinum for catalysts, sapphire for bearings, diamond for cutting tools. They range in price from UK£90,000 (US$200,000) to well over UK£1 m (US$2.2 m) per tonne.

As a good example of how price and availability affect the choice of material for a particular job, let us look at how the materials used for building bridges in Cambridge have changed over the centuries. As our photograph of Queens' Bridge (Fig. 1.2) shows, until 150 years or so ago wood was one of the most common materials for bridge building. It was cheap, and high-quality timber was still available in large sections from natural forests. Stone, too, as the picture of Clare Bridge (Fig. 1.3) shows, was commonly used. In the eighteenth century the availability of large quantities of cast-iron, with its relatively low assembly costs, led to many cast-iron bridges of the type exemplified by Magdalene Bridge (Fig. 1.4). Later metallurgical developments allowed large mild-steel structures to be built from the late nineteenth century on (the Fort St. George Footbridge, Fig. 1.5). Finally, the advent of cheap reinforced concrete led to aesthetically attractive and corrosion-resistant structures like the Garret Hostel Lane bridge (Fig. 1.6). This evolution of building materials clearly illustrates the availability of materials. Nowadays, wood, steel and reinforced concrete are often used interchangeably in structures, reflecting the relatively small price differences between them. The choice of which of the three materials to use is often dictated by complex economic factors which can change from day to day.

Engineering design, then, involves many considerations (Fig. 1.7). The choice of a material must meet certain criteria on bulk and surface properties (strength and

Fig. 1.2. The wooden bridge at Queens' College, a 1904 copy of the original "mathematical" bridge built in 1749 to William Etheridge's design.

Fig. 1.3. Clare Bridge, built in 1640, is Cambridge's oldest surviving bridge; it is reputed to have been an escape-route from the college in times of plague.

Fig. 1.4. Magdalene Bridge built in 1823 on the site of the ancient Saxon bridge over the Cam. The present cast-iron arches carried, until recently, loads far in excess of those envisaged by the designers. Fortunately, the bridge is now undergoing a well-earned restoration.

Fig. 1.5. A typical twentieth-century mild-steel bridge; a convenient crossing to the Fort St. George inn!

Fig. 1.6. The reinforced concrete footbridge in Garret Hostel Lane. An inscription carved nearby reads: "This bridge was given in 1960 by the Trusted family members of Trinity Hall. It was designed by Timothy Guy MORGAN an undergraduate of Jesus College who died in that year."

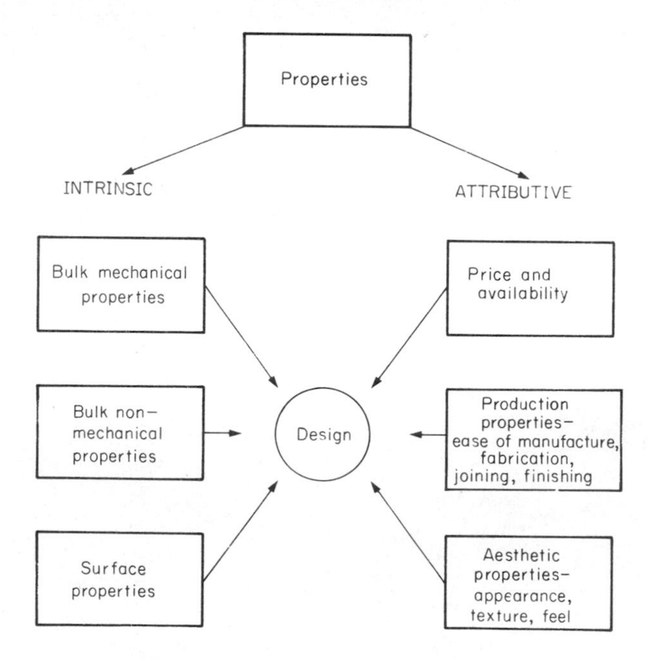

Fig. 1.7. How the properties of engineering materials affect the way in which products are designed.

corrosion-resistance, for example). But it must also be able to be fabricated; it must appeal to potential consumers; and it must compete economically with other alternative materials. In the next chapter we consider the economic aspects of this choice, returning in later chapters to a discussion of the other properties.

Further reading

J. E. Gordon, *The New Science of Strong Materials, or Why You Don't Fall Through the Floor*, Penguin, 1976 (excellent general introduction to materials).

A. Price and availability

CHAPTER 2

THE PRICE AND AVAILABILITY OF MATERIALS

Introduction

In the first chapter we introduced the range of properties required of engineering materials by the design engineer, and the range of materials available to provide these properties. We ended by showing that the *price* and *availability* of materials were important and often overriding factors in selecting the materials for a particular job. In this chapter we examine these economic properties of materials in more detail.

Data for material prices

Because cost is so important in any design exercise, we give a ranked list of current prices of a range of materials in Table 2.1. This is simply an itemised version of Table 1.3 in Chapter 1; and the subdivisions of materials into classes by price that we mentioned in Chapter 1 can be applied just as well to the materials in Table 2.1. Where do such data come from? And how do we keep informed about materials prices and what controls them?

Trade supply journals list current prices. A typical such journal is *Procurement Weekly*, listing current prices of basic materials, together with prices 6 months and a year ago. All manufacturing industries take this weekly—the workshop in your engineering department will have it—and it gives a good guide to prices and their trends. Table 2.2 is abstracted from the prices table of *Procurement Weekly* for January 16, 1980. It shows the part dealing with non-ferrous metals, and illustrates two points. First, there is a long-term upward movement in material prices. Twenty years ago, copper was UK£200 (US$440) per tonne, and lead was UK£110 (US$240) per tonne; now they are more than four times this price. Second, there are large short-term fluctuations in material prices. Lead has dropped 30% in 6 months. Gold has risen 140% in the same period. Nickel has risen 47% in the last 12 months, but in the previous year it fell 40%. These are large changes, important to the purchaser of materials.

The short-term price fluctuations have little to do with the real scarcity or abundance of materials. They are caused by small differences between the rate of supply and demand, much magnified by speculation in commodity futures. The volatile nature of

TABLE 2.1
PRICE PER TONNE, $\tilde{p}$ (JANUARY 1980)

Material	$\tilde{p}$/UK£ tonne^{-1}	$\tilde{p}$/US$ tonne^{-1}
Diamonds, industrial	400,000,000	900,000,000
Platinum	12,000,000	26,000,000
Gold	8,700,000	19,100,000
Silver	520,000	1,140,000
Boron-epoxy composites (mats. 60% of cost; fabr. 40% of cost)	150,000	330,000
CFRP (mats. 30% of cost; fabr. 60% of cost)	90,000	200,000
Cobalt/tungsten carbide cermets	30,000	66,000
Tungsten	11,800	26,000
Cobalt	7800	17,200
Titanium alloys	4630–5780	10,190–12,720
Polyimides	4600	10,100
Nickel	3196	7031
PMMA	2400	5300
High-speed steel	1816	3995
Nylon 66	1495	3289
GFRP (mats. 60% of cost: fabr. 40% of cost)	1100–1500	2400–3300
Stainless steels	1100–1400	2400–3100
Copper, worked (sheet, tubes, bars)	1024–1360	2253–2990
Copper, ingots	1024	2253
Polycarbonate	1160	2550
Aluminium alloys, worked (sheet, bars)	910–1110	2000–2440
Aluminium, ingots	910	2000
Brass, worked (sheet, tubes, bars)	750–1062	1650–2336
Brass, ingots	684	1505
Magnesia, MgO	950	1990
Alumina, Al_2O_3	500–800	1100–1760
Zinc, worked (sheet, tubes, bars)	430–791	950–1740
Zinc, ingots	333	733
Lead, worked (bars, sheet, tube)	500-760	1100–1670
Lead, ingots	437	961
Epoxy	750	1650
Glass	680	1500
Foamed polymers	400–650	880–1430
Natural rubber	650	1430
Polypropylene	580	1280
Polyethylene, high density	570	1250
Polystyrene	605	1330
Hard woods	590	1300
Polyethylene, low density	550	1210
Polyvinyl chloride	360	790
Silicon carbide	200–350	440–770
Plywood	340	750
Low-alloy steels	175–250	385–550
Mild steel, worked (angles, sheet, bars)	200–220	440–480
Cast iron	120	260
Iron, ingots	108	238
Soft woods	196	431
Concrete, reinforced (beams, columns, slabs)	125–135	275–297
Fuel oil	90	200
Coal	38	84
Cement	24	53

TABLE 2.2
SAMPLE OF PRICE DATA FOR JANUARY 16, 1980

	Current	6 months ago	12 months ago
Chromium, lump/UK£ (US$) $tonne^{-1}$	3500 (7700)	3200 (7040)	2900 (6380)
Copper, cathodes/UK£ (US$) $tonne^{-1}$	1052 (2314)	926 (2037)	764 (1680)
Gold/US$ Troy oz^{-1}	610	251	222
Lead, refined pig/UK£ (US$) $tonne^{-1}$	437 (961)	578 (1270)	428 (942)
Mercury/US$ $Flask^{-1}$	390	300	127
Nickel, refined/UK£ (US$) $tonne^{-1}$	3196 (7031)	2675 (5885)	2169 (4772)
Tungsten, powder/UK£ (US$) kg^{-1}	11.8 (26.0)	12.8 (28.2)	13.5 (29.7)

the commodity market can result in large changes over a period of a few days—that is one reason speculators are attracted to it—and there is very little that an engineer can do to foresee or insure against these changes. Political factors are also extremely important—a scarcity of cobalt in 1978 was due to the guerilla attacks on mineworkers in Zaire, the world's principal producer of cobalt.

The long-term changes are of a different kind. They reflect, in part, the real cost (in capital investment, labour and energy) of extracting and transporting the raw material and processing it to give the engineering material. Inflation and increased energy costs obviously drive the price up; so, too, does the necessity to extract materials, like copper, from increasingly lean ores: the leaner the ore, the more machinery and energy are required to crush the rock containing it, and to concentrate it to the level that the metal can be extracted.

In the long term, then, it is important to know which materials are basically plentiful, and which are likely to become scarce. It is also important to know the extent of our dependence on materials.

TABLE 2.3
UK IMPORTS OF ENGINEERING MATERIALS, RAW AND SEMIS
PERCENTAGE OF TOTAL COST

Material	%
Iron and steel	27
Wood and lumber	21
Copper	13
Plastics	9.7
Silver and platinum	6.5
Aluminium	5.4
Rubber	5.1
Nickel	2.7
Zinc	2.4
Lead	2.2
Tin	1.6
Pulp/paper	1.1
Glass	0.8
Tungsten	0.3
Mercury	0.2
Etc.	1.0

The use-pattern of materials

The way in which materials are used in a developed nation is fairly standard. All consume steel, concrete and wood in construction; steel and aluminium in general engineering; copper in electrical conductors; polymers in appliances, and so forth: and roughly in the same proportions.

About 20% of the total import bill of a country like Britain is spent on engineering materials. Table 2.3 shows how this is distributed among materials. Iron and steel, and the raw materials used to make them, accounted for about a quarter of it. Next are wood and lumber—still widely used in light construction. More than a quarter is spent on the metals copper, silver, aluminium and nickel. All polymers taken together, including rubber, account for little more than 10%. If we include the further metals zinc, lead, tin, tungsten and mercury, the list accounts for 99% of all the money spent abroad on materials, and we can safely ignore the contribution of materials which do not appear on it.

Ubiquitous materials

The composition of the earth's crust

Let us now shift attention from what we *use* to what is widely *available*. A few engineering materials are synthesised from compounds found in the earth's oceans and atmosphere: magnesium is an example. But almost all are won by mining their ore from the earth's crust, and concentrating it sufficiently to allow the material to be extracted or synthesised from it. How plentiful and widespread are these materials on which we depend so heavily? How much copper, silver, tungsten, tin and mercury in useful concentrations does the crust contain? All five are rare: workable deposits of them are so small and highly localised that many governments classify them as of strategic importance, and stockpile them.

Not all materials are so thinly spread. Table 2.4 shows the relative abundance of the commoner elements in the earth's crust. The crust is 47% oxygen by weight or—because oxygen is a big atom—96% by volume (geologists are fond of saying that the earth's crust is solid oxygen containing a few percent of impurities). Next in abundance are the elements silicon and aluminium; by far the most plentiful solid materials available to us are silicates and alumino-silicates. A few metals appear on the list, but iron and aluminium are the only ones which appeared also in the list of widely used materials. We have carried the list as far as carbon because it is the backbone of virtually all polymers, including wood. The oceans and the atmosphere show a similar picture. Overall, oxygen and its compounds are overwhelmingly plentiful—on every hand we are surrounded by ceramics, or the raw materials to make them. Some materials are widespread, notably iron and aluminium; but even for these the local concentration is almost always small, usually too small to make it economic to extract them. In fact, the raw materials for making polymers are more readily available at present than those for most metals. There are huge deposits of carbon in the earth. On a world scale, we extract a greater tonnage of carbon every month than we extract iron

TABLE 2.4
ABUNDANCE OF ELEMENTS/WEIGHT PERCENT

Crust		Oceans		Atmosphere	
Oxygen	47	Oxygen	85	Nitrogen	79
Silicon	27	Hydrogen	10	Oxygen	19
Aluminium	8	Chlorine	2	Argon	2
Iron	5	Sodium	1	Carbon dioxide	0.04
Calcium	4	Magnesium	0.1		
Sodium	3	Sulphur	0.1		
Potassium	3	Calcium	0.04		
Magnesium	2	Potassium	0.04		
Titanium	0.4	Bromine	0.007		
Hydrogen	0.1	Carbon	0.002		
Phosphorus	0.1				
Manganese	0.1				
Fluorine	0.06				
Barium	0.04				
Strontium	0.04				
Sulphur	0.03				
Carbon	0.02				

* The total mass of the crust to a depth of 1 km is 3×10^{21} kg; the mass of the oceans is 10^{20} kg; that of the atmosphere is 5×10^{18} kg.

in a year, but at present we simply burn it. And the second ingredient of most polymers—hydrogen—is also one of the most plentiful of elements.

Some materials—iron, aluminium, the elements to make glass and cement—are plentiful and widely available. But others—mercury, silver, tungsten are examples—are scarce and highly localised, and may not last very long.

Exponential growth and consumption doubling-time

How do we calculate the life-time of a resource like mercury? Like almost all materials, mercury is being consumed at a rate which is growing exponentially with time (Fig. 2.1). If the current rate of consumption in tonnes per year is C then exponential growth means that

$$\frac{dC}{dt} = \frac{r}{100} C \tag{2.1}$$

where r is the fractional rate of growth in % per year. Integrating gives

$$C = C_0 \exp\left\{\frac{r(t - t_0)}{100}\right\} \tag{2.2}$$

where C_0 was the consumption rate at time $t = t_0$. The *doubling-time* t_D of consumption is given by setting $C/C_0 = 2$ to give

$$t_D = \frac{100}{r} \log_e 2 \approx \frac{70}{r}. \tag{2.3}$$

Steel consumption is growing at 3.4% per year—it doubles every 20 years roughly. Aluminium consumption is rising at about 8% per year—it doubles every 9 years.

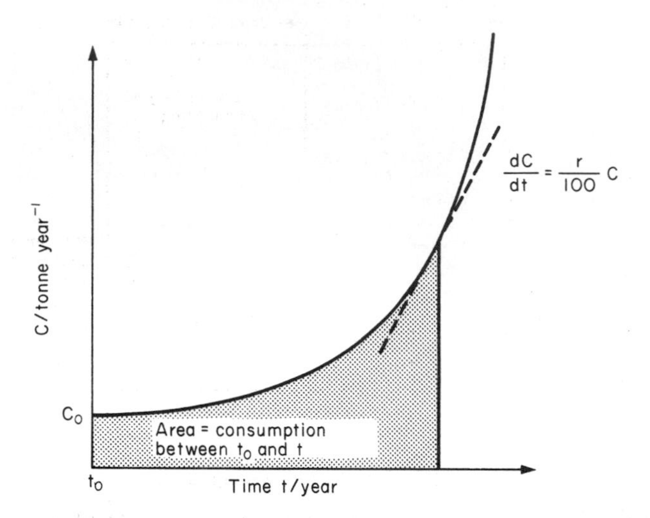

Fig. 2.1. The exponentially rising consumption of materials.

Polymer production in the US has increased at 18% per year over the last few years—it doubles every 4 years.

Resource availability

The availability of a resource depends on the degree to which it is *localised* in one or a few countries (making it susceptible to production controls or cartel action); on the *size* of the reserves, or, more accurately, the resource base (see below); and on the *energy* required to mine and process it. The influence of the last two (size of reserves and energy content) can, within limits, be studied and predicted.

The calculation of resource life involves the important distinction between reserves and resources. The current reserve is the known deposits which can be mined profitably at today's price using today's technology; it bears little relationship to the true magnitude of the resource base; in fact, the two are not even roughly proportional.

The resource base includes the current reserve. But it also includes all deposits that might become available given adequate prospecting and which, by various extrapolation techniques, can be estimated. And it includes, too, all known and unknown deposits that cannot be mined profitably now, but which—due to higher prices, better technology or improved transportation—might reasonably become available in the future (Fig. 2.2). The reserve is like money in the bank—you *know* you have got it. The resource base is more like your total potential earnings over your lifetime—it is much larger than the reserve, but it is less certain, and you may have to work very hard to get it. The resource base is the only realistic measure of the total available material. Resources are almost always much larger than reserves. But because the geophysical data and economic projections are so poor, their evaluation is subject to vast uncertainty.

Although the resource base is uncertain, it obviously is very important for us to estimate how long it can continue to supply a particular material. That for oil is

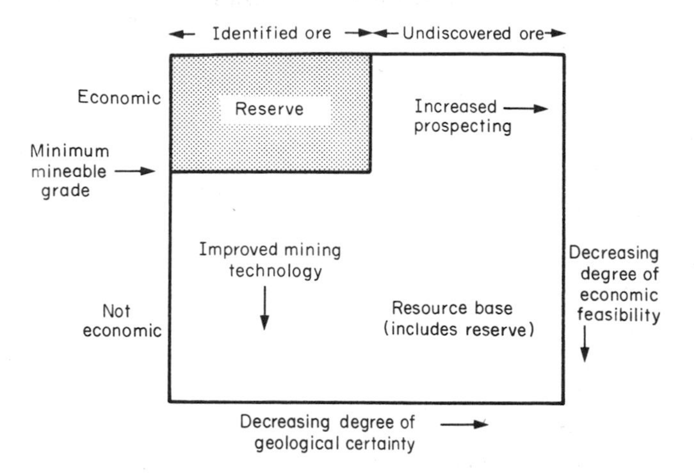

Fig. 2.2. The McElvey diagram.

estimated (assuming that the present use-pattern continues) at about 25 years; that for natural gas only slightly longer. Assuming that demand continues to rise exponentially, our exponential formula allows us to estimate how long it would take us to use up *half* of the resources. (At this stage prices would begin to rise so steeply that supply would become a severe problem.) For a number of important materials these half-lives lie within your life-time: for silver, tin, tungsten, zinc, lead and mercury, for example, they lie between 50 and 80 years. Others, most notably iron, aluminium, and the raw materials from which most ceramics and glasses are made, have enormous resource bases, adequate for hundreds of years, even allowing for continued exponential growth.

The cost of energy enters here in an important way. The extraction of materials requires energy (Table 2.5). As a material becomes scarcer—copper is a good example—we have to extract it from leaner and leaner ores. This expends more and more energy per tonne of copper *metal* in mining, crushing and concentrating the ore, and these energy costs rapidly become prohibitive. The rising energy content of copper shown in Table 2.5 reflects the fact that the richer copper ores are being rapidly worked out at this very moment.

TABLE 2.5
APPROXIMATE ENERGY CONTENT OF MATERIALS GJ TONNE^{-1}

Material	GJ tonne^{-1}
Aluminium	300
Plastics	100
Copper	100, rising to 500
Zinc	70
Steel	50
Glass	20
Cement	8
Brick	4
Timber	2
Gravel	0.1
Oil	44
Coal	29

* Energy costs UK£2 (US$4.4) per GJ in 1980.

The future

How are we going to cope with the shortages of engineering materials in the future? One way obviously is by

Material-efficient design

Many current designs use far more material than is necessary, or use potentially scarce materials where the more plentiful would serve. Often, for example, it is a surface property (e.g. low friction, or high corrosion resistance) which is required—then a thin surface film on a cheap plentiful substrate can replace the bulk use of a scarcer material. Another way of adjusting to shortages is by

Substitution

It is almost always the property, not the material itself, that the user wants. Another, more readily available material can often be used instead, but usually at considerable outlay (new processing methods, new joining methods, etc.). Examples of substitution are the replacement of stone and wood by steel and concrete that we saw in our first chapter, the replacement of copper by polyethylene in plumbing, the change from wood and metals to polymers in household goods, and from copper to aluminium in electrical wiring.

There are, however, technical limitations to substitution. Some materials we use in ways not easily filled by others. Platinum as a catalyst, liquid helium as a refrigerant, and (at present) silver in photography cannot be replaced. Others—a replacement for tungsten for lamp filaments, for example—would require the development of a whole new technology—and this can take many years. Finally, substitution increases the demand for the replacement material, which may also be in limited supply. The massive trend to substitute plastics for other materials (current growth rate about 15% per year) puts a heavier burden on petrochemicals, at present derived from oil. A third approach is that of

Recycling

This is not new: old building materials have been recycled for millennia; scrap metal has been recycled for decades, and is a major industry. Now recycling is labour intensive, not requiring much capital or energy. Over the last 30 years, the rising cost of labour has limited its scope. But if energy and capital become relatively scarcer (as they are doing now), recycling will become much more attractive. There will be an increasing incentive to design manufactured goods so that they can be taken apart more easily and reused.

Conclusion

Overall, the materials-resource problem is not as critical as that of energy. Some materials have an enormous base—and fortunately these include the major structural materials. For others, the resource base is small, but they are often used in small quantities so that the price could rise a lot without having a drastic effect on the price of the product; and for some, substitutes are available. But such adjustments can take up to 25 years if a new technology is needed—and capital. Rising energy costs, plus rising material costs as the Third World gains control of its own resources, mean that the relative costs of materials will change rapidly in the next 20 years, and a good designer must be continually aware of these changes, and continually on the look out for opportunities to substitute one material for another.

Further reading

A. H. Cottrell, *Environmental Economics*, Arnold, 1977.
T. Darwent (ed.), *World Resources—Engineering Solutions*, Inst. Civil Engineers, London, 1976.
E. G. Kovach (ed.), *Technology of Efficient Energy Utilisation*, NATO Science Committee, Brussels, 1973.
E. G. Kovach (ed.), *The Rational Use of Potentially Scarce Metals*, NATO Science Committee, Brussels, 1976.

B. The elastic moduli

CHAPTER 3

THE ELASTIC MODULI

Introduction

The next material property that we shall examine is the *elastic modulus*. The modulus measures the resistance of a material to elastic (or "springy") deformation. Low modulus materials are floppy, and will suffer large deflections if loaded. Sometimes this is desirable, of course: springs, cushions, vaulting poles—these structures are designed to deflect, and the right choice of modulus may be a low one. But in the great majority of mechanical applications, deflection is undesirable, and the engineer seeks a material with a high modulus. If, for example, rods of identical cross-section are subjected to identical loadings, they deflect elastically by very different amounts depending on whether they are made from steel, silica, wood or nylon. The modulus is reflected, too, in the natural frequency of vibration of a structure. A beam of low modulus has a lower natural frequency than one of higher modulus (though the density matters also) and this, as well as the deflection, is important in design calculations.

Before we look in detail at the modulus, we must first define stress and strain.

Definition of stress

Imagine a block of material to which we apply a force F (see Fig. 3.1(a)). The force is transmitted through the block and is balanced by an equal, opposite force that the base exerts on the block (if this were not so, the block would move); so we can replace the base by the equal and opposite force. F acts on sections through the block, parallel to the original surface: the whole of the block is said to be in a state of stress. The intensity of the stress, σ, is measured by the force divided by the area, A, of the block face, giving

$$\sigma = \frac{F}{A}. \tag{3.1}$$

This particular stress is caused by a force pulling at right angles to the face; we call it the *tensile* stress.

Suppose now that the force acted not normal to the face but at an angle to it, as shown in Fig. 3.1(b). We can resolve the force into two components, one normal to the face and the other parallel to it. The normal component causes a state of tensile stress to appear in the block. Its magnitude, as before, is F_t/A.

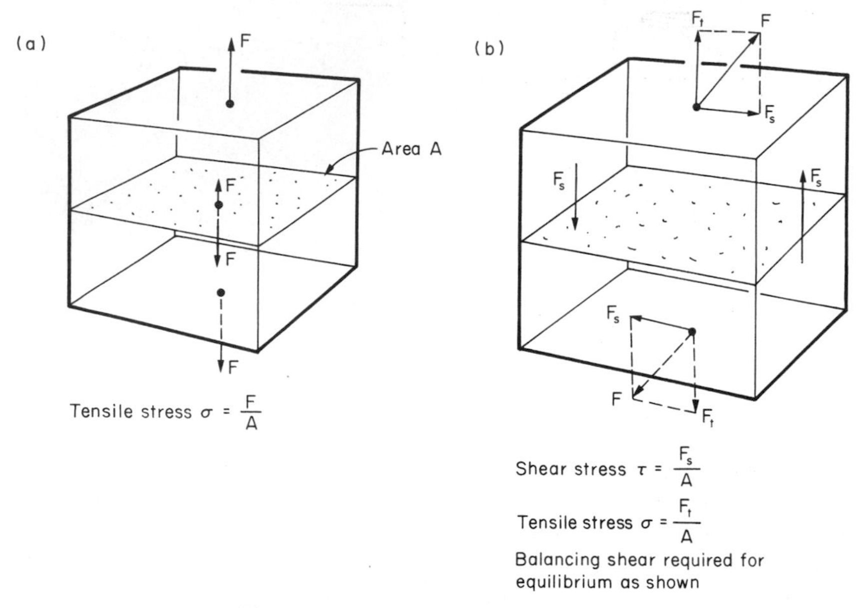

Fig. 3.1. Definitions of stress: σ and τ.

The other component of the force, F_s, also loads the block, but it does so in *shear.* The shear stress, τ, in the block parallel to the direction of F_s is given by

$$\tau = \frac{F_s}{A}. \tag{3.2}$$

The important point is that the magnitude of a stress is always equal to the magnitude of a *force* divided by the *area* of the face on which it acts. Forces are measured in newtons, so stresses are measured in units of newtons per metre squared ($\mathrm{N\,m^{-2}}$). For many engineering applications, this is inconveniently small, and the normal unit of stress is the mega newton per metre squared or mega pascal ($\mathrm{MN\,m^{-2}}$ or MPa).

There are four commonly occurring states of stress, shown in Fig. 3.2. The simplest is that of *simple tension* or *compression* (as in a tension member loaded by pin joints at its ends or in a pillar supporting a structure in compression). The stress is, of course, the force divided by the section area of the member or pillar. The second common state of stress is that of *biaxial tension.* If a spherical shell (like a balloon) contains an internal pressure, then the skin of the shell is loaded in two directions, not one, as shown in Fig. 3.2. This state of stress is called biaxial tension. (Unequal biaxial tension is obviously the state in which the two tensile stresses are unequal.) The third common state of stress is that of *hydrostatic pressure.* This occurs deep in the earth's crust, or deep in the ocean, when a solid is subjected to equal compression on all sides. There is a convention that stresses are *positive* when they *pull,* as we have drawn them in earlier figures. Pressure, however, is positive when it *pushes,* so that the magnitude of the pressure differs from the magnitude of the other stresses in its sign. Otherwise it is defined in exactly the same way as the force divided by the area on which it acts. The final common state of stress is that of *pure shear.* If you try to twist a thin tube, then

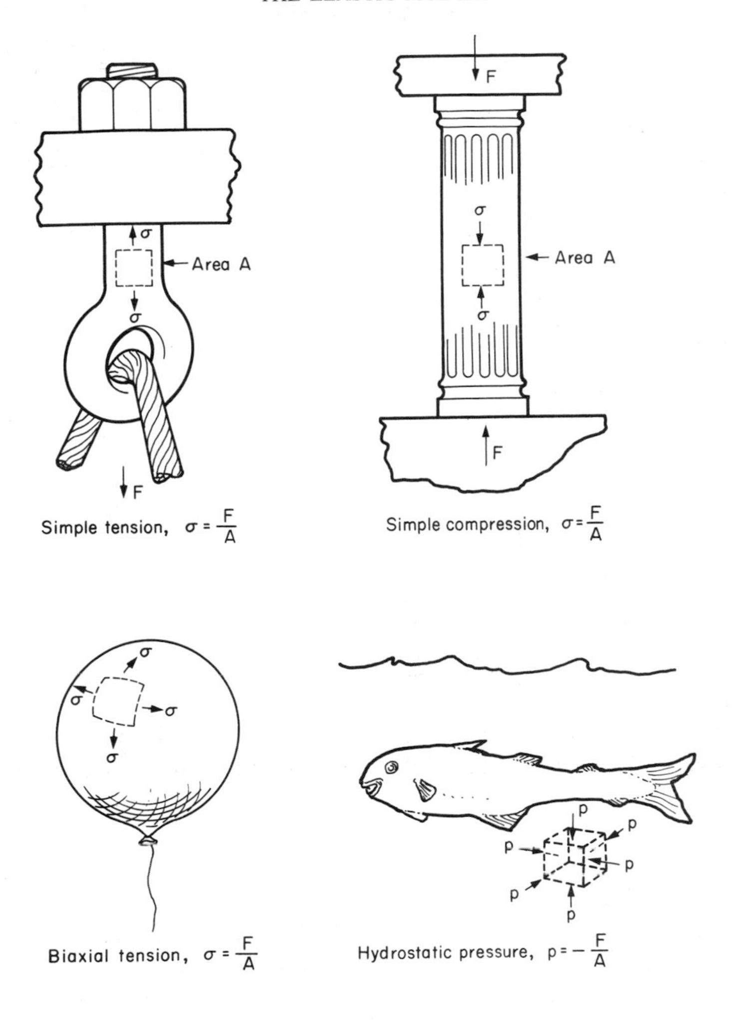

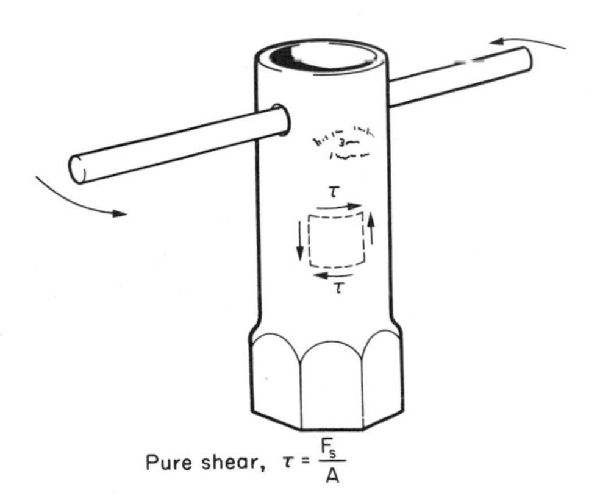

Fig. 3.2. Common states of stress.

elements of it are subjected to pure shear, as shown. This shear stress is simply the shearing force divided by the area of the face on which it acts.

Remember one final thing; if you know the stress in a component, then the *force* acting across any face of it is the stress times the area.

Strain

Materials respond to stress by *straining*. Under a given stress, a stiff material (like steel) strains only slightly; a floppy or compliant material (like polyethylene) strains much more. The modulus of the material describes this property, but before we can measure it, or even define it, we must define strain properly.

The kind of stress that we called a tensile stress induces a tensile strain. If the stressed cube of side l, shown in Fig. 3.3, extends by an amount u parallel to the

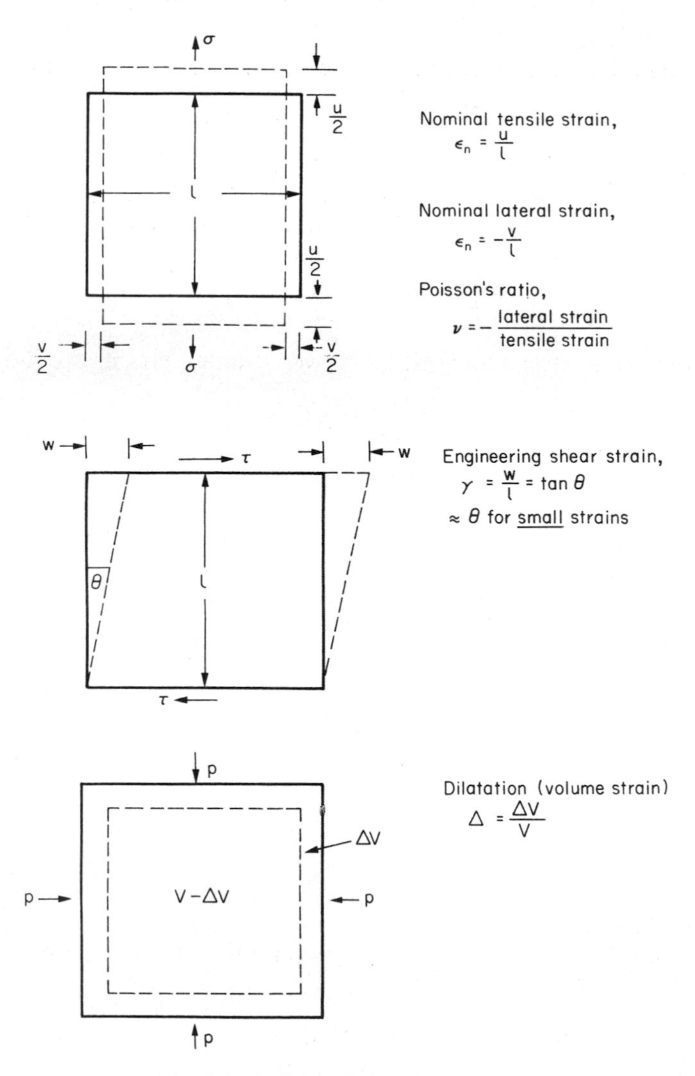

Fig. 3.3. Definition of strain: ε_n, γ, Δ.

tensile stress, we define the *nominal tensile strain* as

$$\varepsilon_n = \frac{u}{l}. \tag{3.3}$$

When it strains in this way, the cube usually gets thinner. The amount by which it shrinks inwards is described by Poisson's ratio, ν, which is the negative of the ratio of the inward strain to the original tensile strain:

$$\text{lateral strain} = -\nu \text{ tensile strain.}$$

A shear stress induces a shear strain. If a cube shears sideways by an amount w then the *shear strain* is defined by

$$\gamma = \frac{w}{l} = \tan \theta \tag{3.4}$$

where θ is the angle of shear. Since the elastic strains are almost always very small, we may write, to a good approximation,

$$\gamma = \theta.$$

Finally, hydrostatic pressure induces a volume change called dilatation. If the volume change is ΔV and the cube volume is V, we define the *dilatation* by

$$\Delta = \frac{\Delta V}{V}. \tag{3.5}$$

Since they are the ratios of two lengths, or of two volumes, strains are dimensionless.

Hooke's Law

We can now define the elastic moduli. They are defined through Hooke's Law, which is merely a description of the experimental observation that, when strains are small, the strain is very nearly proportional to the stress for many materials.

The nominal tensile strain, for example, is proportional to the tensile stress; for *simple* tension

$$\sigma = E\varepsilon_n, \tag{3.6}$$

where E is called *Young's modulus*. The same relationship also holds for stresses and strains in simple compression, of course.

Secondly, the shear strain is proportional to the shear stress, with

$$\tau = G\gamma, \tag{3.7}$$

where G is the *shear modulus*.

Finally, the negative of the dilatation is proportional to the pressure (because positive pressure causes a shrinkage of volume) with

$$p = -K\Delta, \tag{3.8}$$

where K is called the *bulk modulus*.

Because strain is dimensionless, the moduli have the same dimensions as those of stress: force per unit area ($N\,m^{-2}$). In practice, it is convenient to measure the moduli in units of giga newtons per metre squared, a giga newton being 10^9 newtons ($GN\,m^{-2}$ or GPa).

This linear relationship between stress and strain is a very handy one when calculating the response of a *linear elastic solid* to stress, but it must be remembered that most solids are elastic only to *very small* strains: up to about 0.001. Beyond that some break and some become plastic—and this we will discuss in later chapters. A few solids like rubber are elastic up to very much larger strains of order 4 or 5, but they cease to be *linearly* elastic (that is the stress is no longer proportional to the strain) after a strain of about 0.01.

One final point. We earlier defined Poisson's ratio as the negative of the lateral shrinkage strain to the tensile strain. This quantity, Poisson's ratio, is an elastic constant, so we have four elastic constants that crop up commonly: E, G, K and ν. In a moment when we discuss data for these constants we list data only for E. For most metals it is useful to know, however, that

$$K \approx E, \; G \approx \tfrac{3}{8}E \text{ and } \nu \approx 0.33, \qquad (3.9)$$

although for other materials, the relationship can be much more complicated.

Measurement of Young's modulus

How do we measure Young's modulus for a material? A simple way is to compress it by applying a known compressive force to a block of the material, and measure the strain. Young's modulus is then given by $E = \sigma/\varepsilon_n$.

But in general, this is a poor way of measuring the modulus. For one thing, if the modulus is large, the extension u may be too small to measure with precision. And, for another, if anything else contributes to the strain, like creep (which we will discuss in a later chapter), then it will obviously lead to an incorrect value for E.

An example of a much better way of measuring E is to measure the natural frequency of vibration of a rod of the material simply supported at its ends (Fig. 3.4) and heavily loaded by a mass M at the middle (so that we may neglect the mass of the rod

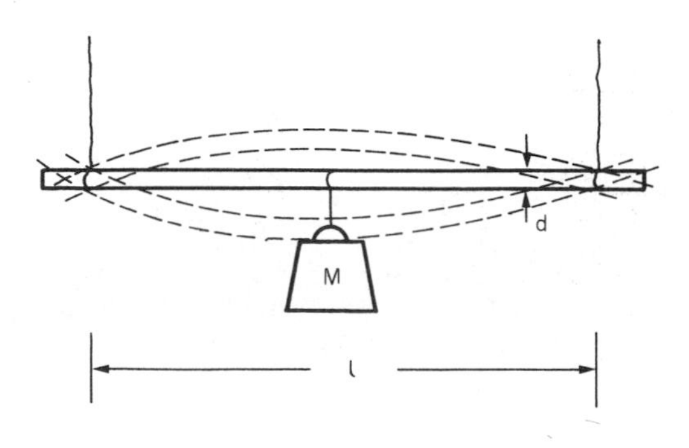

Fig. 3.4.

itself). The frequency of oscillation of the rod, f cycles per second (or hertz), is given by

$$f=\frac{1}{2\pi}\left\{\frac{3\pi Ed^4}{4l^3M}\right\}^{1/2} \tag{3.10}$$

from which

$$E=\frac{16\pi Ml^3f^2}{3d^4}. \tag{3.11}$$

Use of stroboscopic techniques and carefully designed apparatus can make this sort of method very accurate.

TABLE 3.1
DATA FOR YOUNG'S MODULUS, E

Material	E/GN m^{-2}	Material	E/GN m^{-2}
Diamond	1000	Niobium and alloys	80–110
Tungsten carbide, WC	450–650	Silicon	107
Osmium	551	Zirconium and alloys	96
Cobalt/tungsten carbide cermets	400–530	Silica glass, SiO_2 (quartz)	94
Borides of Ti, Zr, Hf	500	Zinc and alloys	43–96
Silicon carbide, SiC	450	Gold	82
Boron	441	Calcite (marble, limestone)	81
Tungsten	406	Aluminium	69
Alumina, Al_2O_3	390	Aluminium and alloys	69–79
Beryllia, BeO	380	Silver	76
Titanium carbide, TiC	379	Soda glass	69
Molybdenum and alloys	320–365	Alkali halides (NaCl, LiF, etc.)	15–68
Tantalum carbide, TaC		Granite (Westerly granite)	62
Niobium carbide, NbC		Tin and alloys	41–53
Silicon nitride, Si_3N_4		Concrete, cement	45–50
Chromium	289	Fibreglass (glass-fibre/epoxy)	35–45
Beryllium and alloys	200–289	Magnesium and alloys	41–45
Magnesia, MgO	250	GFRP	7–45
Cobalt and alloys	200–248	Calcite (marble, limestone)	31
Zirconia, ZrO	160–241	Graphite	27
Nickel	214	Alkyds	20
Nickel alloys	130–234	Shale (oil shale)	18
CFRP	70–200	Common woods, ‖ to grain	9–16
Iron	196	Lead and alloys	14
Iron-based superalloys	193–214	Ice, H_2O	9.1
Ferritic steels, low-alloy steels	200–207	Melamines	6 7
Stainless austenitic steels	190–200	Polyimides	3–5
Mild steel	196	Polyesters	1–5
Cast irons	170–190	Acrylics	1.6–3.4
Tantalum and alloys	150–186	Nylon	2–4
Platinum	172	PMMA	3.4
Uranium	172	Polystyrene	3–3.4
Boron/epoxy composites	125	Polycarbonate	2.6
Copper	124	Epoxies	3
Copper alloys	120–150	Common woods, ⊥ to grain	0.6–1.0
Mullite	145	Polypropylene	0.9
Zircon, ZrO_2	145	Polyethylene, high density	0.7
Vanadium	130	Foamed polyurethane	0.01–0.06
Titanium	116	Polyethylene, low density	0.2
Titanium alloys	80–130	Rubbers	0.01–0.1
Palladium	124	PVC	0.003–0.01
Brasses and bronzes	103–124	Foamed polymers	0.001–0.01

The best of all methods of measuring E is to measure the velocity of sound in the material. The velocity of longitudinal waves, v_l, depends on Young's modulus and the density, ρ.

$$v_l = \left(\frac{E}{\rho}\right)^{1/2}. \tag{3.12}$$

v_l is measured by "striking" one end of a bar of the material (by glueing a piezo-electric crystal there and applying a charge-difference to the crystal surfaces) and measuring the time sound takes to reach the other end (by attaching a second piezo-electric crystal). Most moduli are measured by one of these last two methods.

Data for Young's modulus

Now we can look at the values for Young's moduli of materials. Table 3.1 is a ranked list of moduli of materials which will be useful in solving problems and in selecting materials for particular applications. Diamond is at the top of the list, with a modulus of 10^3 GN m^{-2}; and soft rubbers and various foamed polymers are at the bottom with moduli as low as 10^{-3} GN m^{-2}. You can, of course, make special materials with lower

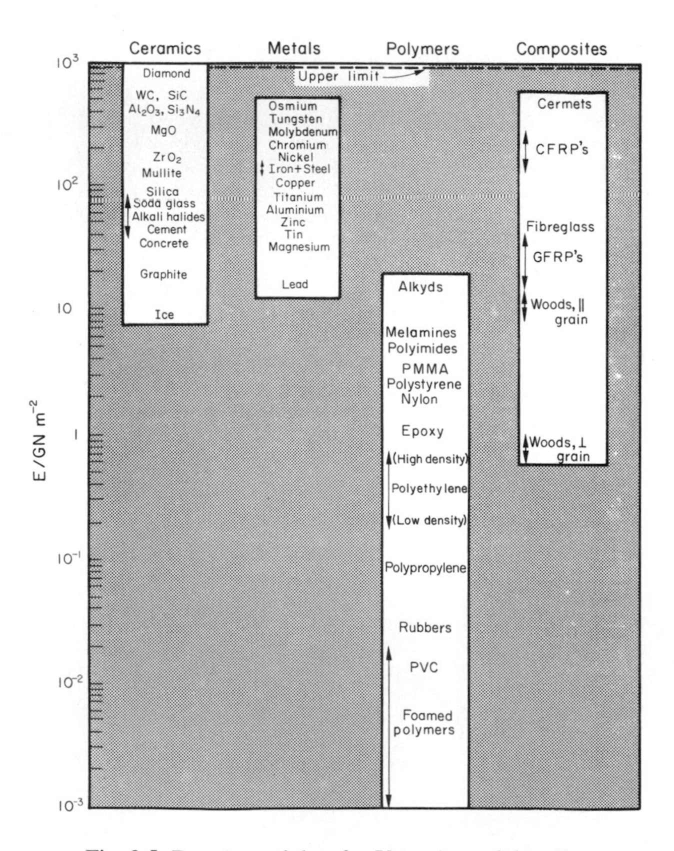

Fig. 3.5. Bar-chart of data for Young's modulus, E.

moduli—jelly, for instance, has a modulus of about 10^{-6} GN m^{-2}! But practical engineering materials lie in the range 10^{-3} to 10^{+3} GN m^{-2}—a range of 10^6. This is the range you have to choose from when selecting a material for a given application.

A good perspective of the moduli of materials is given by the bar-chart shown in Fig. 3.5. It shows the usual range of moduli from 10^{-3} to 10^{+3} GN m^{-2}. Ceramics and metals—even the floppiest of them, like lead—lie near the top of this range. Polymers are much more compliant, the common ones (polyethylene, PVC and polypropylene) lying several decades lower.

To understand the origin of the modulus, why it has the values it does, why polymers are much less stiff than metals, and what we can do about it, we have to examine the *structure* of materials, and *the nature of the forces* holding the atoms together. In the next two chapters we will examine this, and then return to the modulus, and to our bar-chart, with new understanding.

Further reading

A. H. Cottrell, *The Mechanical Properties of Matter*, Wiley, 1964, Chap. 4.
S. P. Timoshenko and J. N. Goodier, *Theory of Elasticity*, McGraw Hill, 1970, Chap. 1.
C. J. Smithells, *Metals Reference Book*, 5th edition, Butterworths, 1976 (for data).

CHAPTER 4

BONDING BETWEEN ATOMS

Introduction

In order to understand material properties like *Young's modulus*, we need to take a close look at materials at the *atomic* level. Two things are especially important in influencing the modulus:

(1) The forces which hold atoms together (the *interatomic bonds*) which act like little springs, linking one atom to the next in the solid state (Fig. 4.1)

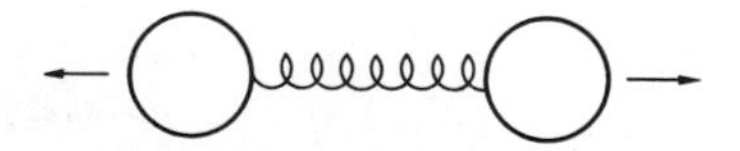

Fig. 4.1.

and

(2) The ways in which atoms pack together (the *atom packing*), since this determines how many little springs there are per unit area, and the angle at which they are pulled (Fig. 4.2).

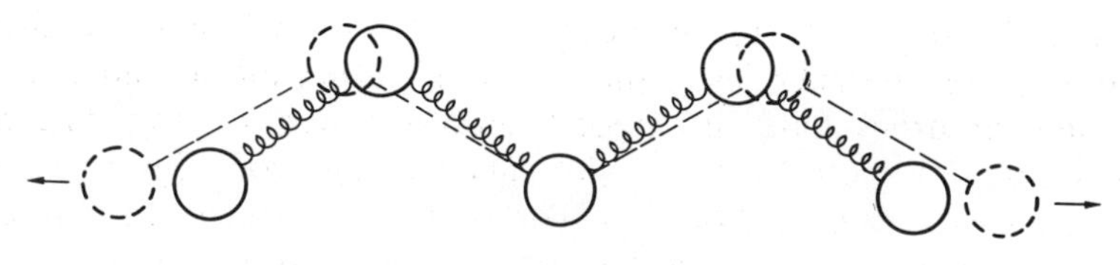

Fig. 4.2.

In this chapter we shall look at the forces which can bind atoms together to give *bonds*, and in the next at the ways in which atoms can be packed together.

The various ways in which atoms can be bound together involve

(1) **Primary bonds**—*ionic*, *covalent* or *metallic* bonds, which are all relatively *strong* (they generally melt between 1000 and 5000 K), and
(2) **Secondary bonds**—*Van der Waals* and *hydrogen bonds*, which are both relatively *weak* (they melt between 100 and 500 K).

We should remember, however, when drawing up a list of distinct bond types like this that many atoms are really bound together by bonds which contain more than one type of bond (mixed bonds).

Primary bonds

Ceramics and metals are entirely held together by primary bonds—the ionic and covalent bond in ceramics, and the metallic and covalent bond in metals. These strong, stiff bonds give high moduli.

The ionic bond is typified by cohesion in sodium chloride. Other alkali halides (such as lithium fluoride), oxides (magnesia, alumina) and components of cement (hydrated carbonates and oxides) are wholly or partly held together by ionic bonds.

Let us start with the *sodium atom*. This consists of a nucleus of 11 *protons* (and 12 neutrons) surrounded by 11 *electrons* (Fig. 4.3):

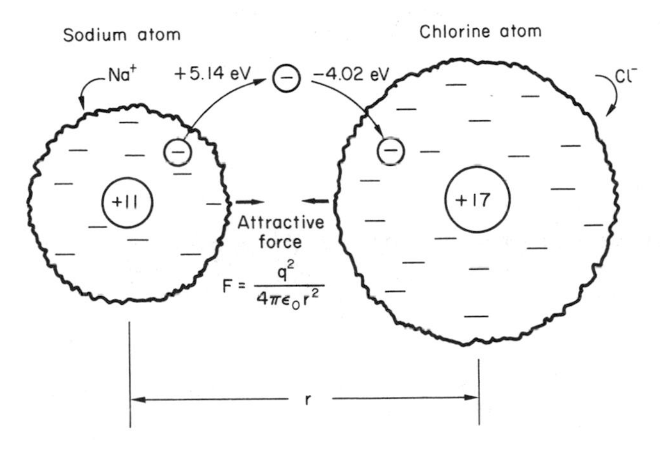

Fig. 4.3. The formation of an ionic bond—in this case between a sodium atom and a chlorine atom, making sodium chloride.

The electrons are attracted to the nucleus by electrostatic forces and therefore have negative energies. But the energies of the electrons are not all the same. Those furthest from the nucleus naturally have the highest (least negative) energy. The electron that we can most easily remove from the sodium atom is therefore the outermost one: we can remove it by expending 5.14 eV of work. This electron can be most profitably transferred to a vacant position on a distant chlorine atom, giving us back 4.02 eV of energy. Thus, we can make isolated Na^+ and Cl^- by doing 5.14 eV − 4.02 eV = 1.12 eV of work, U_i.

So far, we have had to *do* work to create the ions which will make the ionic bond: it does not seem to be a very good start. However, the + and − charges attract each other, and if we now bring them together, the force of attraction does work. This force is simply that between two opposite point charges:

$$F = q^2/4\pi\varepsilon_0 r^2 \tag{4.1}$$

where q is the charge on each ion, ε_0 is the permittivity of vacuum, and r is the separation of the ions. The work done as the ions are brought to a separation r (from infinity) is:

$$U = \int_r^\infty F\,\mathrm{d}r = q^2/4\pi\varepsilon_0 r. \tag{4.2}$$

Figure 4.4 shows us how we gain more and more work as r decreases, until, at $r \approx 1$ nm for a typical ionic bond, we have paid off the 1.12 eV of work borrowed to form Na^+ and Cl^- in the first place. For $r < 1$ nm, it is all gain, and our ionic bond now becomes more and more stable.

Why does not r decrease indefinitely, releasing more and more energy, and ending in the *fusion* of the two ions? Well, when the ions get close enough together, the electronic charge distributions start to overlap one another, and this causes a very large repulsion. Figure 4.4 shows the potential energy increase that this causes. Clearly, the ionic bond is most stable at the minimum in the $U(r)$ curve, which is well approximated by

$$U = U_i - \underbrace{\frac{q^2}{4\pi\varepsilon_0 r}}_{\text{attractive part}} + \underbrace{\frac{B}{r^n}}_{\text{repulsive part}}. \tag{4.3}$$

How much can we bend this bond? Well, the electrons of each ion occupy complicated three-dimensional regions (or "orbitals") around the nuclei. But at an approximate level we can assume the ions to be spherical, and there is then considerable freedom in the way we pack the ions round each other. The ionic bond therefore *lacks directionality*, although in packing ions of opposite sign, it is obviously necessary to make sure that the total charge (+ and −) adds up to zero.

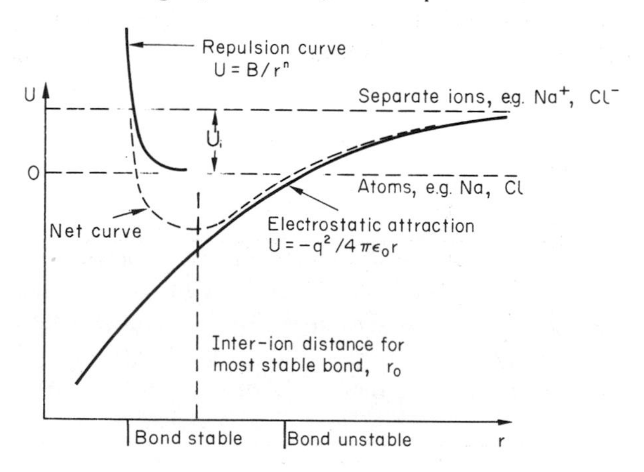

Fig. 4.4. The formation of an ionic bond, viewed in terms of energy.

Covalent bonding appears in its pure form in diamond, silicon and germanium—all materials with large moduli (that of diamond is the highest known). It is the dominant bond-type in silicate ceramics and glasses (stone, pottery, brick, all common glasses, components of cement) and contributes to the bonding of the high-melting-point metals (tungsten, molybdenum, tantalum, etc.). It appears, too, in polymers, linking carbon atoms to each other along the polymer chain; but because polymers also contain bonds of other, much weaker, types (see below) their moduli are usually small.

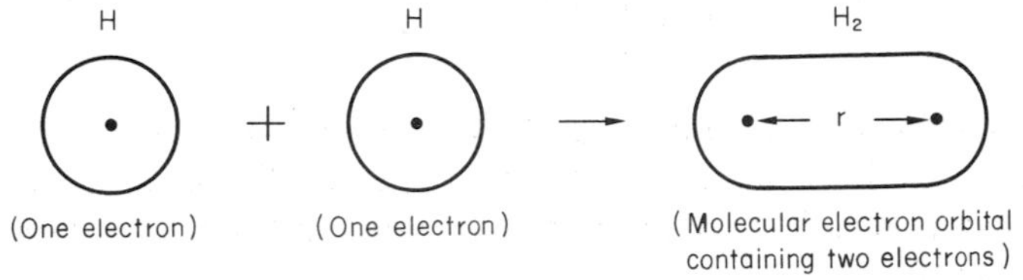

H + H → H_2

(One electron) (One electron) (Molecular electron orbital containing two electrons)

Fig. 4.5. The formation of a covalent bond—in this case between two hydrogen atoms, making a hydrogen *molecule.*

The simplest example of covalent bonding is the hydrogen molecule. The proximity of the two nuclei creates a new electron orbital, shared by the two atoms, into which the two electrons go (Fig. 4.5). This sharing of electrons leads to a reduction in energy, and a stable bond, as Fig. 4.6 shows. The energy of a covalent bond is well described by the empirical equation

$$U = \underbrace{-\frac{A}{r^m}}_{\text{attractive part}} + \underbrace{\frac{B}{r^n}}_{\text{repulsive part}} \qquad (m < n). \tag{4.4}$$

Hydrogen is hardly an engineering material. A more relevant example of the covalent bond is that of diamond, a solid form of carbon used widely for rock-drilling bits. Here, the shared electrons occupy regions that point to the corners of a tetrahedron, as shown in Fig. 4.7(a). The highly symmetrical shape of these orbitals leads to a very *directional* form of bonding in diamond, as Fig. 4.7(b) shows. Depending on orbital shape, many other covalent bonds show various kinds of directionality which, in turn, determines how the atoms pack together to form crystals (Chapter 5).

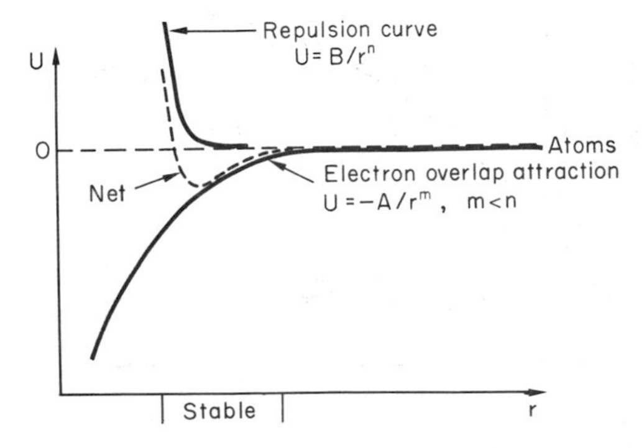

Fig. 4.6. The formation of a covalent bond, viewed in terms of energy.

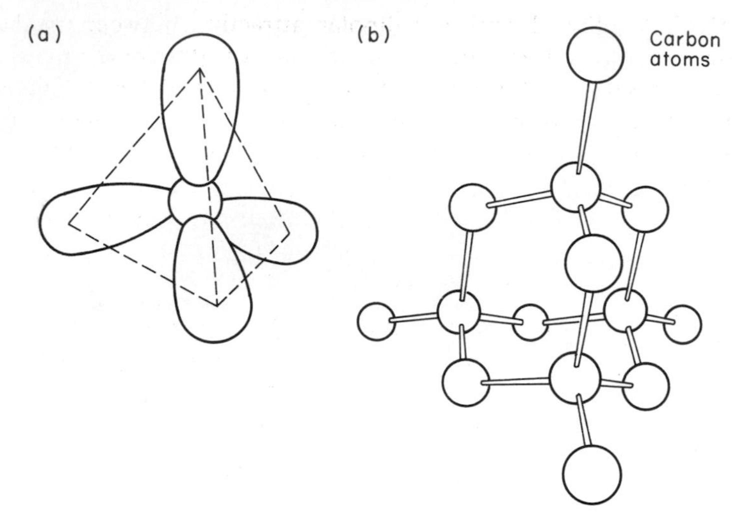

Fig. 4.7. Directional covalent bonding in diamond.

The metallic bond, as the name says, is the dominant (though not the only) bond in metals and their alloys. In a solid (or, for that matter, a liquid) metal, the highest energy electrons tend to leave the parent atoms (making the latter ions) and combine to form a "sea" of rather freely wandering electrons (Fig. 4.8):

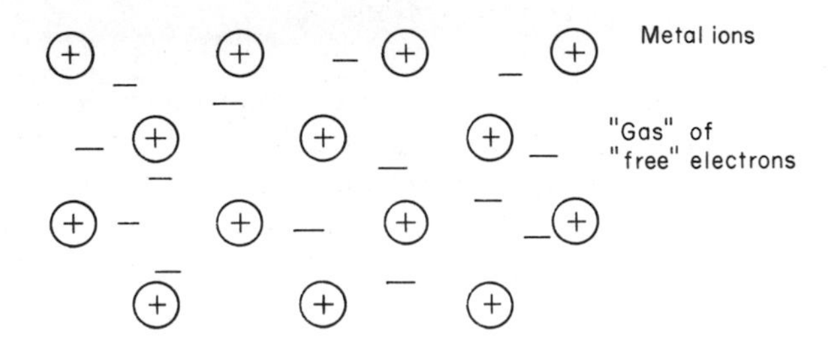

Fig. 4.8. Bonding in a metal—metallic bonding.

The type of energy curve that this gives is very similar to that for covalent bonding, and it is well described by eqn. (4.4).

The easy movement of the electrons leads to the high electrical conductivity of metals. The metallic bond has no directionality, so that metal ions tend to pack to give simple, high-density structures, like ball-bearings shaken down in a box.

Secondary bonds

Although much weaker than primary bonds, secondary bonds are still very important. They provide the links between polymer molecules in polyethylene (and other polymers) which make them solids. Without them, water would boil at −80°C, and life as we know it on earth would not exist.

Van der Waals bonding describes a dipolar attraction between *uncharged* atoms. The charge on an atom is in motion; at any instant it is distributed in an unsymmetric way relative to the nucleus (though, averaged over time, it has spherical symmetry). The instantaneous distribution has a dipole moment; this moment induces a like moment on a nearby atom and the two dipoles attract (Fig. 4.9). Dipoles attract such

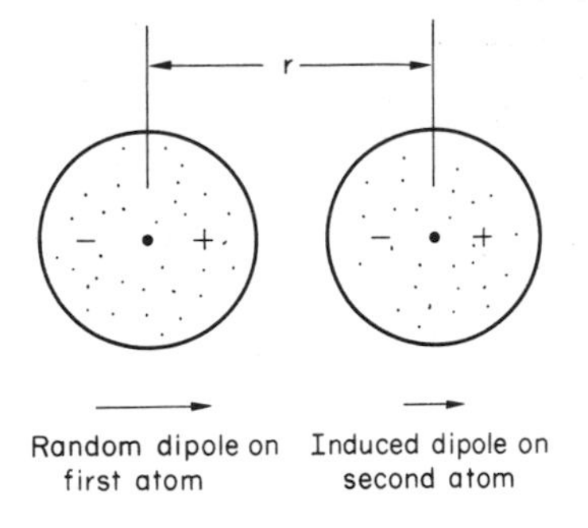

Fig. 4.9. Van der Waals bonding.

that their energy varies as $1/r^6$. Thus the energy of the Van der Waals bond has the form

$$U = \underbrace{-\frac{A}{r^6}}_{\text{attractive part}} + \underbrace{\frac{B}{r^n}}_{\text{repulsive part}} \qquad (n \approx 12). \tag{4.5}$$

A good example is liquid nitrogen, which is kept liquid at −198°C by Van der Waals forces between the covalently bonded N_2 molecules. The thermal agitation produced at room temperature when liquid nitrogen is poured on the floor is ample to break the Van der Waals bonds, showing how weak they are. But without these bonds, most gases would not liquefy, and we should not be able to separate industrial gases from the atmosphere.

Hydrogen bonds keep water liquid at room temperature, and bind polymer chains together to give solid polymers. Ice (Fig. 4.10) is hydrogen-bonded. Each hydrogen

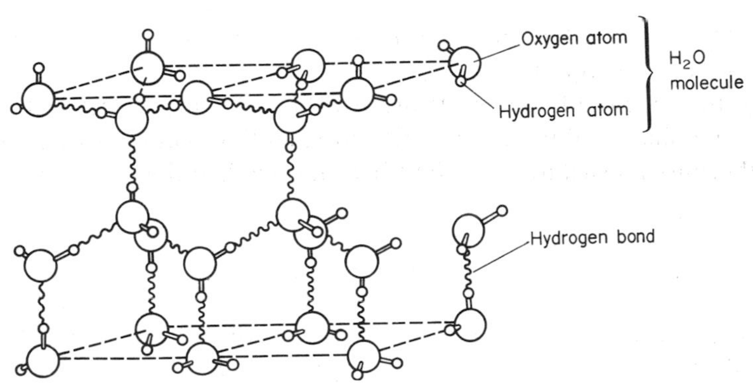

Fig. 4.10. The arrangement of H_2O molecules in ice *I*, showing the hydrogen bonds. The hydrogen bonds keep the molecules well apart, which is why ice *I* has a lower density than water.

atom gives up its charge to the nearest oxygen atom (which then acquires a negative charge). The positively charged H atom (really a proton, since it has lost its electron) acts as a bridging bond between neighbouring oxygen ions, partly because the charge redistribution gives each H_2O molecule a dipole moment (which attracts other H_2O dipoles).

The condensed states of matter

Because these primary and secondary bonds can form, matter in the gaseous state tends to condense to give liquids and solids. Five distinct *condensed states of matter*, differing in their structure and the state of their bonding, can be identified (Table 4.1). The bonds in ordinary liquids have melted, and for this reason the liquid resists compression, but not shear; the bulk modulus, K, is large (compared to the gas) but the shear modulus, G, is zero. The other states of matter, listed in Table 4.1, are distinguished by the state of their bonding (molten versus solid) and their structure (crystalline versus non-crystalline). As Table 4.1 shows, they can also be distinguished by the relative magnitudes of their bulk modulus and shear modulus.

TABLE 4.1
CONDENSED STATES OF MATTER

State	Bonds		Moduli	
	Molten	Solid	K	G and E
1. Liquids	*		Large	Zero
2. Liquid crystals	*		Large	Some non-zero but very small
3. Rubbers	* (2^{ary})	* (1^{ary})	Large	Small ($E \ll K$)
4. Glasses		*	Large	Large ($E \approx K$)
5. Crystals		*	Large	Large ($E \approx K$)

Interatomic forces

Having established the various types of bonds that can form between atoms, and the shapes of their potential energy curves, we are now in a position to find out about the *forces* between atoms. Starting with the $U(r)$ curve, we can find this force F for any separation of the atoms, r, from the relationship

$$F = \frac{dU}{dr}. \tag{4.6}$$

Figure 4.11 shows the shape of the *force*/distance curve that we get from a typical energy/distance curve in this way. Points to note are:

(1) F is zero at the equilibrium separation r_0; however, if the atoms are pulled apart by distance $(r - r_0)$, a force appears so as to resist this pulling apart. The force is closely proportional to $(r - r_0)$ for *small* $(r - r_0)$ for all materials, in both tension and compression.

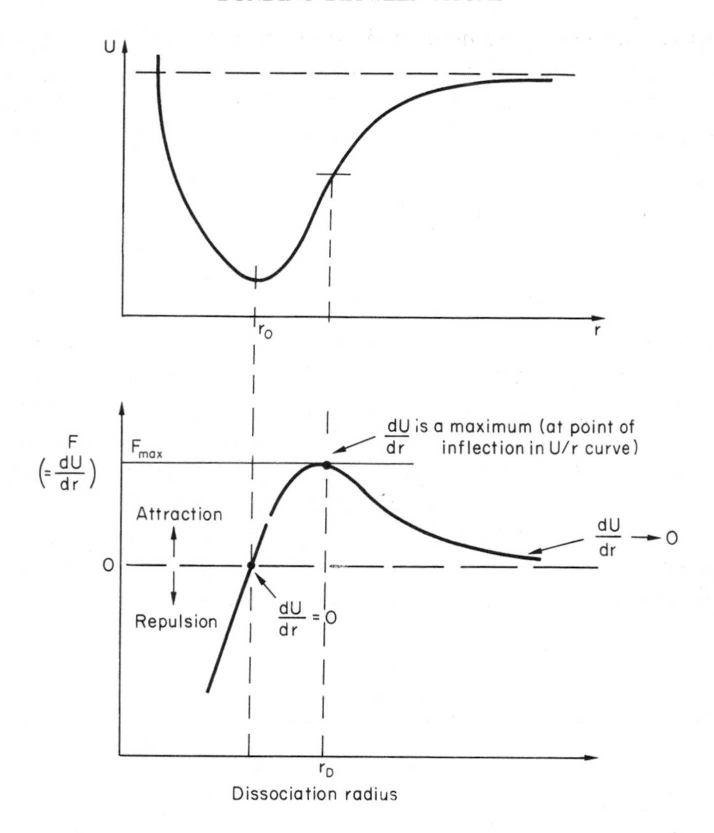

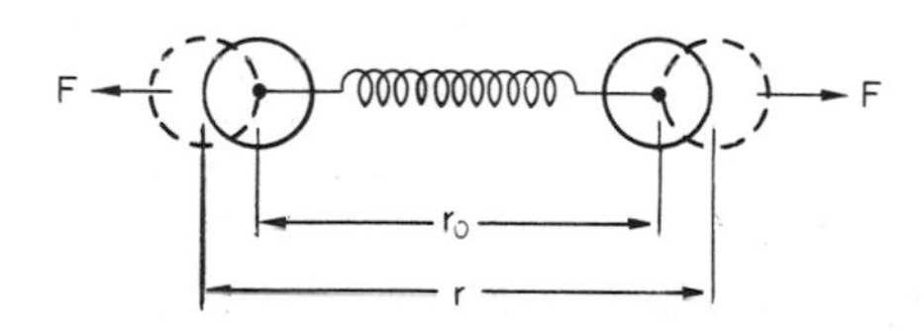

Fig. 4.11.

(2) The *stiffness*, S, of the bond is given by

$$S = \frac{dF}{dr} = \frac{d^2U}{dr^2}. \tag{4.7}$$

When the stretching is *small*, S is constant and equal to

$$S_0 = \left(\frac{d^2U}{dr^2}\right)_{r=r_0} \tag{4.8}$$

that is, the bond behaves in a linear-elastic manner. As we shall see in Chapter 6, this is the physical origin of Hooke's Law.

To conclude, the concept of bond stiffness, based on the energy/distance curves for the various bond types, gets us a long way towards understanding the origin of the modulus. But we need to find out how individual atom bonds build up to form whole pieces of material before we can fully explain experimental data for the modulus. All the types of bonds we have mentioned strongly influence the *packing* of atoms in engineering materials. This is the subject of the next chapter.

Further reading

A. H. Cottrell, *The Mechanical Properties of Matter*, Wiley, 1964, Chap. 2.

K. J. Pascoe, *An Introduction to the Properties of Engineering Materials*, 3rd edition, Van Nostrand, 1978, Chaps. 2, 4.

C. Kittel, *Introduction to Solid State Physics*, 4th edition, Wiley, 1971, Chap. 3.

CHAPTER 5

PACKING OF ATOMS IN SOLIDS

Introduction

In the last chapter, as a first step in understanding the stiffness of solids, we examined the stiffnesses of the bonds holding atoms together. The way in which atoms are packed together is equally important in determining the mechanical properties of our materials. In this chapter we shall show how atoms are arranged in some typical engineering solids.

Atomic packing in crystals

Many engineering materials (almost all metals and ceramics, for instance) are made up entirely of small crystals in which atoms are packed in regular, repeating, three-dimensional patterns. The simpler of these crystals can be understood if the atoms are thought of as *hard spheres* (although, from what we said in the last chapter, it should be obvious that this is a considerable, although convenient, simplification). To make things even simpler, let us for the moment consider a material which is *pure*—with only one size of hard sphere to consider—and which also has *non-directional bonding*, so that we can arrange the spheres subject only to geometrical constraints. Pure copper is a good example of a material satisfying these conditions.

In order to build up a three-dimensional packing pattern, it is easier, conceptually, to begin by

(i) packing atoms two-dimensionally in *atomic planes*,
(ii) stacking these planes on top of one another to give *crystals*.

Close-packed structures and crystal energies

A good example of how we might pack atoms in a *plane*, as in Fig. 5.1, is the arrangement in which the reds are set up on a billiard table before starting a game of snooker. The balls are packed (in a triangular fashion) so as to take up the least possible space on the table. This type of plane is thus called a *close-packed plane*, and contains three types of *close-packed direction*. The figure naturally shows only a small region of close-packed plane—if we had more reds we could extend the plane sideways

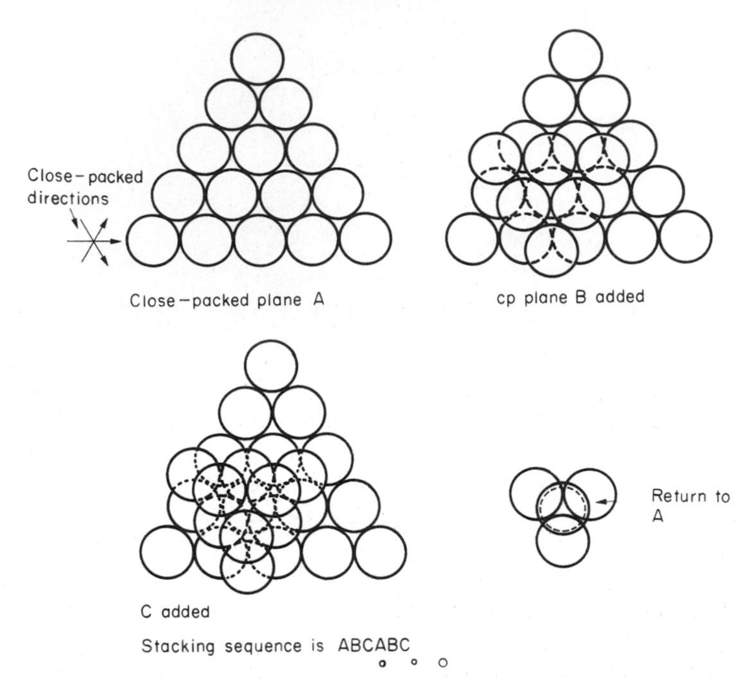

Fig. 5.1. Close packing of "hard-sphere" atoms.

and could, if we wished, fill the whole billiard table. The important thing to notice is the way in which the balls are packed in a *regularly repeating two-dimensional pattern.*

How would we go about adding another layer of atoms to our close-packed plane? As Fig. 5.1 shows, the depressions where the atoms meet are ideal "seats" for the next layer of atoms. By dropping atoms into alternate seats, we can generate a second close-packed plane lying on top of the original one and having an identical packing pattern. Then a third layer can be added, and a fourth, and so on until we have made a sizeable piece of crystal—with, this time, a *regularly repeating pattern of atoms in three dimensions.* The particular structure we have produced is one in which the atoms take up the *least volume* and is therefore called a *close-packed structure.* The atoms in many solid metals are packed in this way.

There is a complication to this apparently simple story. There are two alternative and different *sequences* in which we can stack the close-packed planes on top of one another. If we follow the stacking sequence in Fig. 5.1 rather more closely, we see that, by the time we have reached the *fourth* atomic plane, we are placing the atoms directly above the original atoms (although, naturally, separated from them by the two interleaving planes of atoms). We then carry on adding atoms as before, generating an ABCABC... sequence. In Fig. 5.2 we show the alternative way of stacking, in which the atoms in the *third* plane are now directly above those in the first layer. This gives an ABAB... sequence. These two different stacking sequences give two different three-dimensional packing structures—*Face-centred Cubic* (*f.c.c.*) and *Close-packed Hexagonal* (*c.p.h.*) respectively. Many common metals (e.g. Al, Cu and Ni) have the f.c.c. structure and many others (e.g. Mg, Zn and Ti) have the c.p.h. structure.

Why should Al choose a f.c.c. structure and Mg a c.p.h. structure? The answer is that the f.c.c. structure is the one that gives an Al crystal the least *energy*, and the c.p.h.

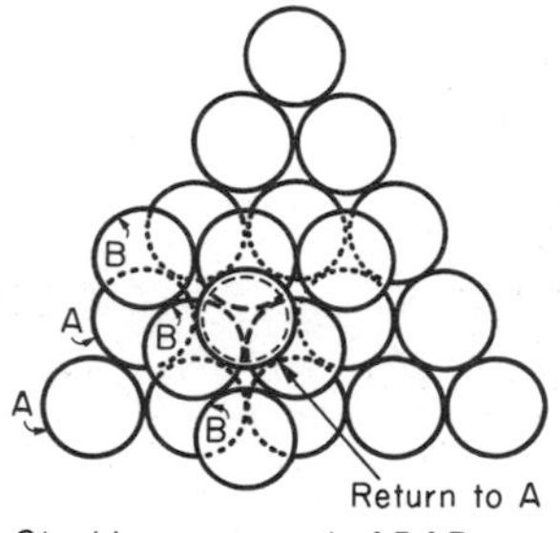

Fig. 5.2. Close packing of "hard-sphere" atoms—an alternative arrangement.

structure the one that gives a Mg crystal the least energy. In general, any material chooses a crystal structure that gives minimum energy, and this structure may not necessarily be close-packed or, indeed, very simple geometrically (although it must still have, of course, some sort of three-dimensional repeating pattern).

The differences in energy between alternative structures is often slight. Because of this, the crystal structure which gives the minimum energy at one temperature may not do so at another. Thus tin changes its crystal structure if it is cooled enough; and, incidentally, becomes much more brittle in the process (causing the tin-alloy coat-buttons of Napoleon's army to fall apart during the harsh Russian winter; and the soldered cans of paraffin on Scott's South Pole expedition to leak with such disastrous consequences). Cobalt changes its structure at 450°C, transforming from an h.c.p. structure at lower temperatures to an f.c.c. structure at higher. More important, pure iron transforms from a b.c.c. structure (defined below) to an f.c.c. structure at 911°C, a process which is important in the heat-treatment of steels.

Crystallography

We have not yet explained why an ABCABC sequence is termed a "f.c.c." structure or why an ABAB sequence is termed a "c.p.h." structure. And we have certainly not attempted to draw planes and stacking sequences for any remotely complicated crystal structure in the way that we were able to do fairly easily for the close-packed structures. In order to explain things such as the geometric differences between f.c.c. and c.p.h. or to ease the conceptual labour of constructing complicated crystal structures, we need an appropriate descriptive language. The method of *crystallography* provides this language, and gives us also an essential shorthand way of describing crystal structures.

Let us illustrate the crystallographic approach in the case of f.c.c. Figure 5.3 shows that the *atom centres* in f.c.c. can be placed at the corners of a cube and in the centres of the cube faces. The cube, of course, has no physical significance but is merely a constructional device. It is called a *unit cell.* If we look along the cube diagonal, we shall see the view shown in Fig. 5.3. A triangular pattern is at once apparent, and only a little effort is needed to see that we are looking at bits of close-packed planes stacked in an ABCABC sequence. This unit-cell visualisation of the atomic positions is thus exactly equivalent to our earlier approach based on stacking of close-packed planes, but is much more powerful as a descriptive aid. For example, we can see how our

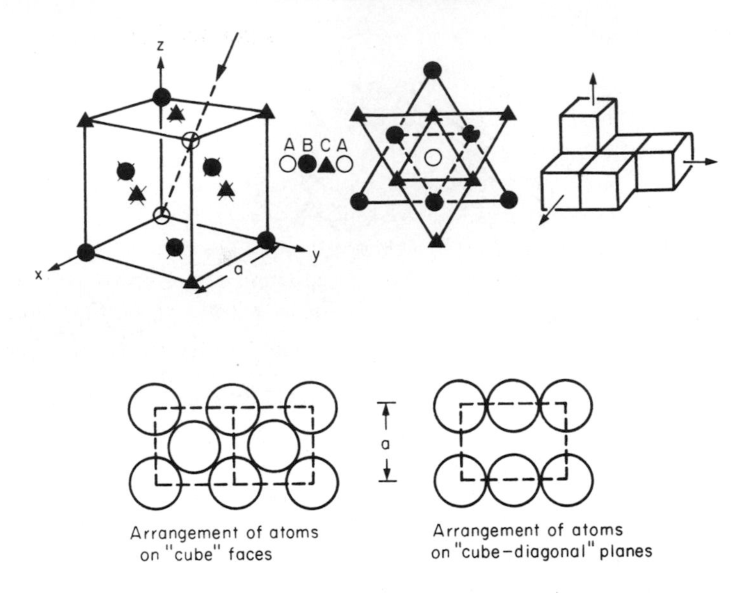

Fig. 5.3. The face-centred-cubic (f.c.c.) structure.

complete f.c.c. crystal is built up by attaching further unit cells to our starting cell rather like assembling a set of children's building cubes. Also, inspection of the unit cell quickly reveals planes in which the atoms are packed other than in a close-packed way. On the "cube" faces the atoms are packed in a square array, and on the cube-diagonal planes in separated rows, as shown in Fig. 5.3. Obviously, properties like the shear modulus might well be different for close-packed planes and cube planes, and it is thus important to have a method of describing various planar packing arrangements.

Let us now look at the c.p.h unit cell as shown in Fig. 5.4. A view looking down the vertical axis reveals the ABA stacking of close-packed planes. We build up our c.p.h. crystal by adding hexagonal building blocks to one another, and, again, can use the unit cell concept to "open up" our view of the various types of planes.

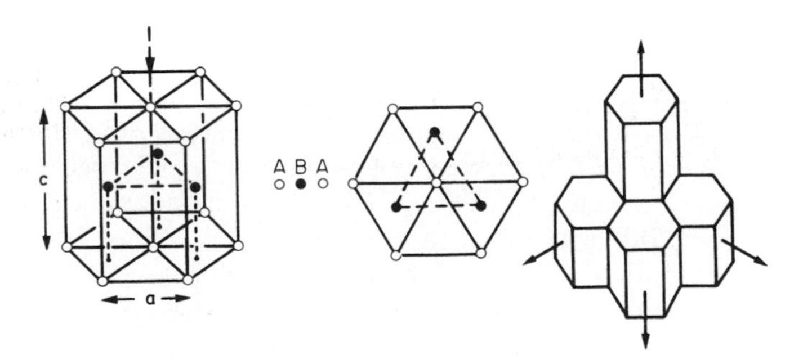

Fig. 5.4. The close-packed-hexagonal (c.p.h.) structure.

Plane indices

We could make scale drawings of the many types of planes that we see in all unit cells; but the concept of a unit cell also allows us to describe any plane by a set of three numbers called *Miller Indices.* The two examples given in Fig. 5.5 should enable you to find the Miller index of any plane in a cubic unit cell. The indices (for a plane) are the *reciprocals* of the intercepts the plane makes with the three axes, reduced to the smallest integers. For example, our six individual "*cube*" *planes* are called (100), (010), (001), together with three parallel planes having identical indices when seen from three adjacent unit cells. Collectively this type of plane is called {100}, with curly brackets. Similarly the six cube *diagonal* planes are (110), ($1\bar{1}0$), (101), ($\bar{1}01$), (011) and ($0\bar{1}1$), or, collectively, {110}. (Here the sign $\bar{1}$ means an intercept of −1.) As a final example, our original close-packed planes are of {111} type. Obviously the unique structural description of "{111} f.c.c." is a good deal more succinct than a scale drawing of close-packed billiard balls!

Different indices are used in hexagonal cells (we build a c.p.h. crystal up by adding bricks in four directions, not three as in cubic) and we will not mention them in this book.

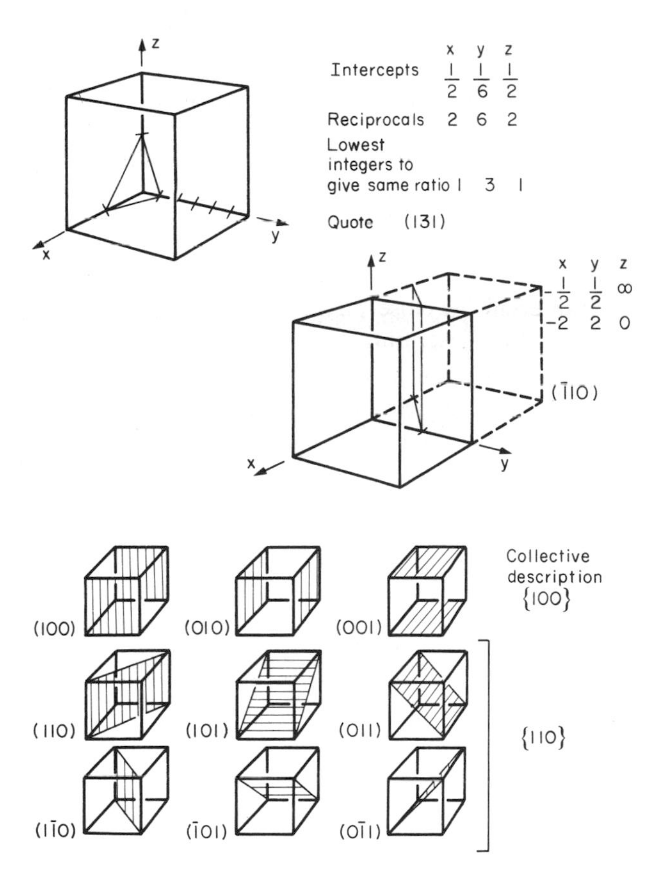

Fig. 5.5. Miller indices for identifying crystal planes.

Direction indices

Properties like Young's modulus may well vary with *direction* in the unit cell, and it is, equally, necessary to have a succinct description of directions. Figure 5.6 shows the method used; and illustrates some typical directions. The indices of direction are the components of a vector (*not* reciprocals), starting from the origin, along the direction, again reduced to the smallest integer set.

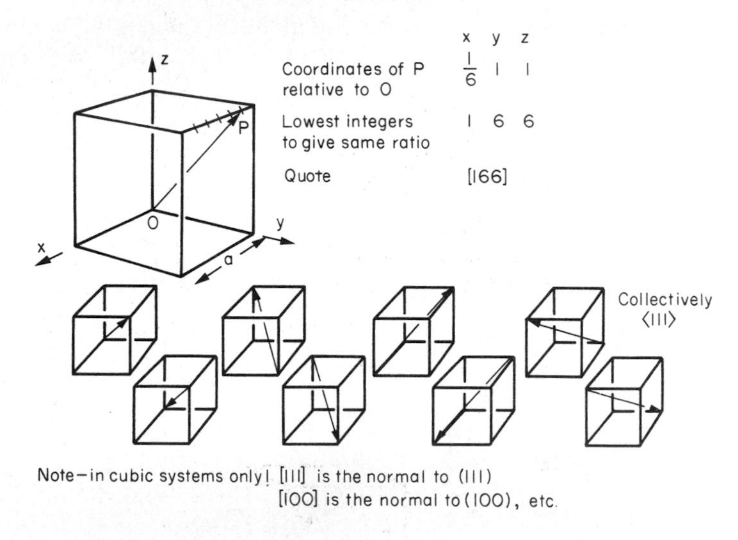

Fig. 5.6. Direction indices for identifying crystal directions.

Other simple, important, crystal structures

Figure 5.7 shows an important example: the body-centred cubic (b.c.c.) structure. The ⟨111⟩ directions are close-packed but there are no close-packed planes. The result is that b.c.c. packing is less dense than either f.c.c. or h.c.p. It is the structure of tungsten, chromium, and iron below 911°C.

In many materials—in the ceramic sodium chloride, for instance—there are *two* species of atoms packed together. Nevertheless, the crystal structures of such compounds can still be simple ones. Figure 5.8(a) shows that the *ceramics* NaCl, KCl and

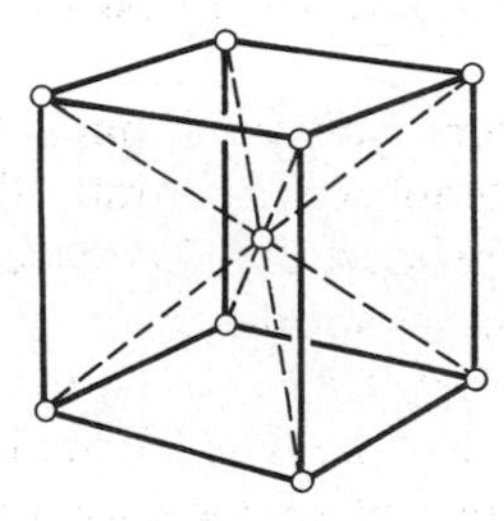

Fig. 5.7. The body-centred-cubic (b.c.c.) structure.

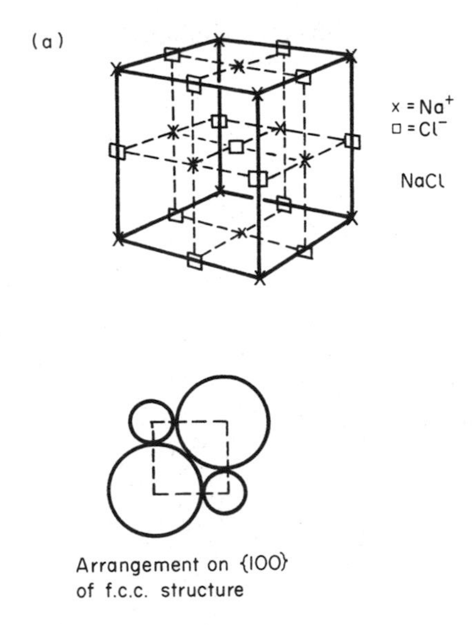

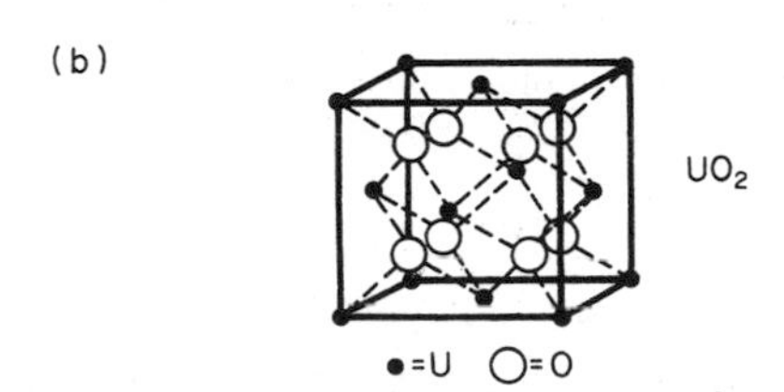

Fig. 5.8. (a) Packing of the unequally sized ions of sodium chloride to give a f.c.c. structure; KCl and MgO pack in the same way. (b) Packing of ions in uranium dioxide; this is more complicated than packing in NaCl because the U and O ions are not in a 1:1 ratio.

MgO, for example, form a cubic structure. Naturally, when two species of atoms are not in the ratio 1:1, as in compounds like the nuclear fuel UO_2 (a ceramic too), the structure is more complicated, as shown in Fig. 5.8(b).

Atomic packing in polymers

As we saw in the first chapter, polymers have become important engineering materials. They are much more complex structurally than metals, and because of this they have very special mechanical properties. The extreme elasticity of a rubber band is one; the formability of polyethylene is another.

Polymers are huge chain-like molecules in which the atoms forming the backbone of the chain are linked by *covalent* bonds. The chain backbone is usually made from carbon atoms (though a limited range of silicon-based polymers can be synthesised). A typical high polymer ("high" means "of large molecular weight") is polyethylene. It is made by the catalytic polymerisation of ethylene:

$$\begin{array}{ccccccccccccccc}
H & & H & & & & H & & H & & H & & H & & H & & H & \\
| & & | & & & & | & & | & & | & & | & & | & & | & \\
C & = & C & \rightarrow & - & & C & - & C & - & C & - & C & - & C & - & C & - \quad \text{etc.} \\
| & & | & & & & | & & | & & | & & | & & | & & | & \\
H & & H & & & & H & & H & & H & & H & & H & & H &
\end{array}$$

Polystyrene, similarly, is made by the polymerisation of styrene:

$$\begin{array}{ccccccccccccccc}
H & & C_6H_5 & & & H & & C_6H_5 & & H & & H & & H & & C_6H_5 & \\
| & & | & & & | & & | & & | & & | & & | & & | & \\
C & = & C & \rightarrow & - & C & - & C & - & C & - & C & - & C & - & C & - \quad \text{etc.} \\
| & & | & & & | & & | & & | & & | & & | & & | & \\
H & & H & & & H & & H & & H & & C_6H_5 & & H & & H &
\end{array}$$

A co-polymer is made by polymerisation of two monomers, adding them randomly (a random copolymer) or in an ordered way (a block copolymer). An example is styrene–butadiene rubber, SBR:

$$\begin{array}{cccccccccccccccccccccccc}
H & & C_6H_5 & & H & & H & & H & & H & & & H & & C_6H_5 & & H & & H & & H & & H & \\
| & & | & & | & & | & & | & & | & & & | & & | & & | & & | & & | & & | & \\
C & = & C & + & C & = & C & - & C & = & C & \rightarrow & - & C & - & C & - & C & - & C & = & C & - & C & - \quad \text{etc.} \\
| & & | & & | & & & & & & | & & & | & & | & & | & & & & & & | & \\
H & & H & & H & & & & & & H & & & H & & H & & H & & & & & & H &
\end{array}$$

Molecules such as these form long, flexible, spaghetti-like chains (Fig. 5.9). Figure 5.10 shows how they pack to form bulk material. The important thing is that, in many polymers, the chains are arranged *randomly* and *not* in regularly repeating three-dimensional patterns. These polymers are thus *non-crystalline*, or *amorphous*. In other polymers the chains can be folded carefully backwards and forwards over one another

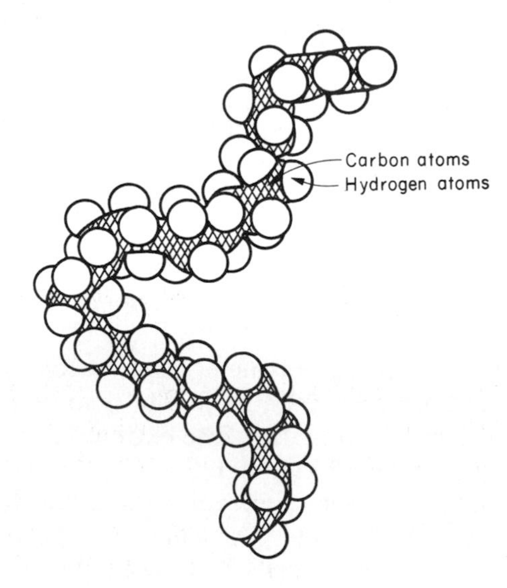

Fig. 5.9. The three-dimensional appearance of a short bit of a polyethylene molecule.

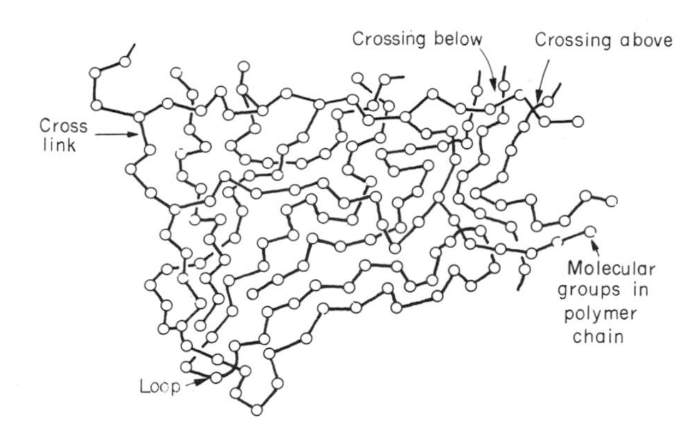

(a) A rubber above its glass-transition temperature. The structure is entirely amorphous. The chains are held together only by occasional covalent cross-linking.

(b) A rubber below its glass-transition temperature. In addition to occasional covalent cross-linking the molecular groups in the polymer chains attract by Van der Waals bonding, tieing the chains closely to one another.

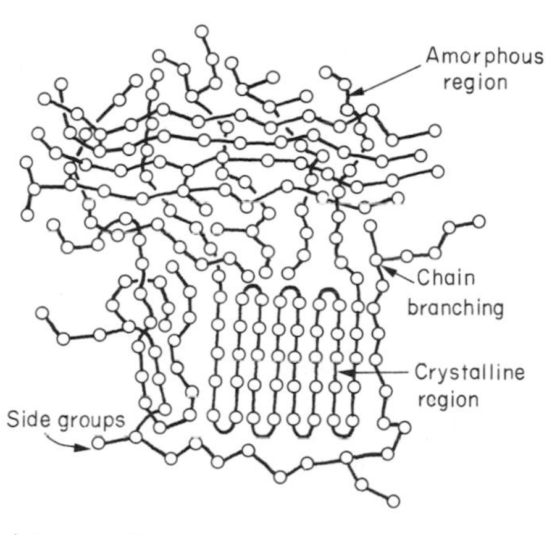

(c) Low-density polyethylene, showing both amorphous and crystalline regions.

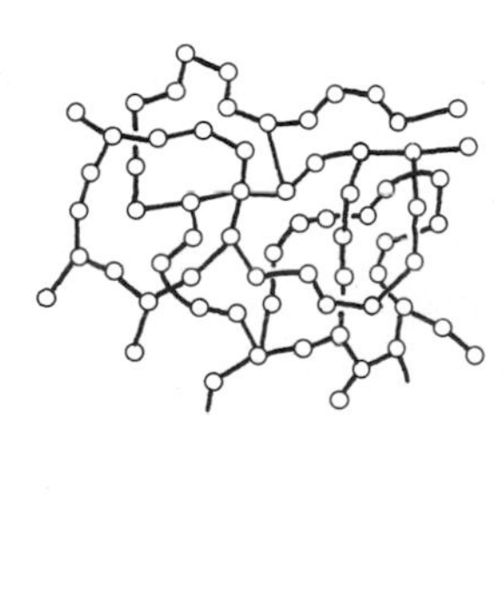

(d) A polymer (e.g. epoxy resin) where the chains are tied tightly together by frequent covalent cross-links.

Fig. 5.10. How the molecules are packed together in polymers.

so as to look like the firework called the "jumping jack". The regularly repeating symmetry of this chain-folding leads to *crystallinity*, giving a crystalline polymer. Finally, some polymers can contain both amorphous and crystalline regions, as shown in Fig. 5.10.

There is a whole science called *molecular architecture* devoted to making all sorts of chains and trying to arrange them in all sorts of ways to make the final material. There are currently thousands of different polymeric materials all having different properties—and new ones are being made every day!

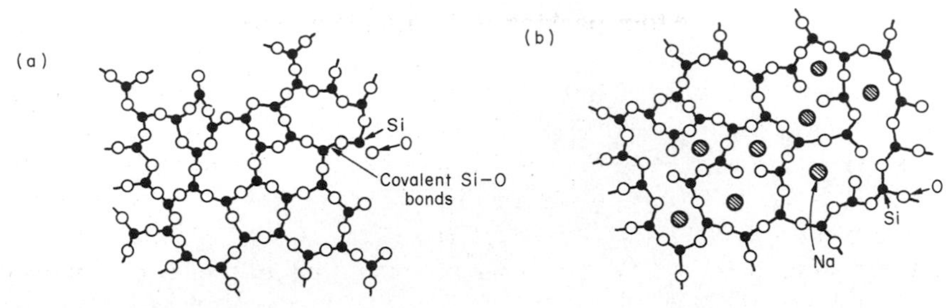

Fig. 5.11. (a) Atom packing in amorphous (glassy) silica. (b) How the addition of soda breaks up the bonding in amorphous silica, giving soda glass.

TABLE 5.1
DATA FOR DENSITY, ρ

Material	ρ/Mg m^{-3}	Material	ρ/Mg m^{-3}
Osmium	22.7	Silicon carbide, SiC	2.5–3.2
Platinum	21.4	Silicon nitride, Si_3N_4	3.2
Tungsten and alloys	13.4–19.6	Mullite	3.2
Gold	19.3	Beryllia, BeO	3.0
Uranium	18.9	Common rocks	2.2–3.0
Tungsten carbide, WC	14.0–17.0	Calcite (marble, limestone)	2.7
Tantalum and alloys	16.6–16.9	Aluminium	2.7
Molybdenum and alloys	10.0–13.7	Aluminium alloys	2.6–2.9
Cobalt/tungsten-carbide cermets	11.0–12.5	Silica glass, SiO_2 (quartz)	2.6
Lead and alloys	10.7–11.3	Soda glass	2.5
Silver	10.5	Concrete/cement	2.4–2.5
Niobium and alloys	7.9–10.5	GFRPs	1.4–2.2
Nickel	8.9	Carbon fibres	2.2
Nickel alloys	7.8–9.2	PTFE	2.3
Cobalt and alloys	8.1–9.1	Boron fibre/epoxy	2.0
Copper	8.9	Beryllium and alloys	1.8–2.1
Copper alloys	7.5–9.0	Graphite, high strength	1.8
Brasses and bronzes	7.2–8.9	Fibreglass (GFRP/Polyester)	1.8
Iron	7.9	PVC	1.3–1.6
Iron-based superalloys	7.9–8.3	CFRPs	1.5–1.6
Stainless steels, austenitic	7.5–8.1	Polyesters	1.1–1.5
Tin and alloys	7.3–8.0	Polyimides	1.4
Low-alloy steels	7.8	Epoxies	1.1–1.4
Mild steel	7.8	Polyurethane	1.1–1.3
Stainless steel, ferritic	7.5–7.7	Polycarbonate	1.2–1.3
Cast iron	6.9–7.8	PMMA	1.2
Titanium carbide, TiC	7.2	Nylon	1.1–1.2
Zinc and alloys	5.2–7.2	Polystyrene	1.0–1.1
Chromium	7.2	Polyethylene, high-density	0.94–0.97
Zirconium carbide, ZrC	6.6	Ice, H_2O	0.92
Zirconium and alloys	6.6	Natural rubber	0.83–0.91
Titanium	4.5	Polyethylene, low-density	0.91
Titanium alloys	4.3–5.1	Polypropylene	0.88–0.91
Alumina, Al_2O_3	3.9	Common woods	0.4–0.8
Alkali halides	3.1–3.6	Foamed plastics	0.01–0.6
Magnesia, MgO	3.5	Foamed polyurethane	0.06–0.2

Atom packing in inorganic glasses

Glasses usually consist of oxides (e.g. SiO_2) with the atoms packed in a non-crystalline (or amorphous) way. Figure 5.11(a) shows schematically the structure of silica glass, which is solid to well over 1000°C because of the strong covalent bonds linking the Si to the O atoms. Addition of soda (Na_2O) breaks up the structure and lowers the *softening temperature* (at which the glass can be worked) to about 700°C. This soda glass (Fig. 5.11(b)) is the material of which milk bottles and window panes are made.

The density of solids

The densities of common engineering materials are listed in Table 5.1 and shown in Fig. 5.12. These reflect the mass and diameter of the atoms that make them up, and the efficiency with which they are packed to fill space in the crystal or glass. Metals, most of them, have high densities because the atoms are heavy and closely packed. Polymers, and many ceramics, are much less dense because the atoms of which they are made (C, H, O) are light, and because they generally adopt structures which are not closely packed.

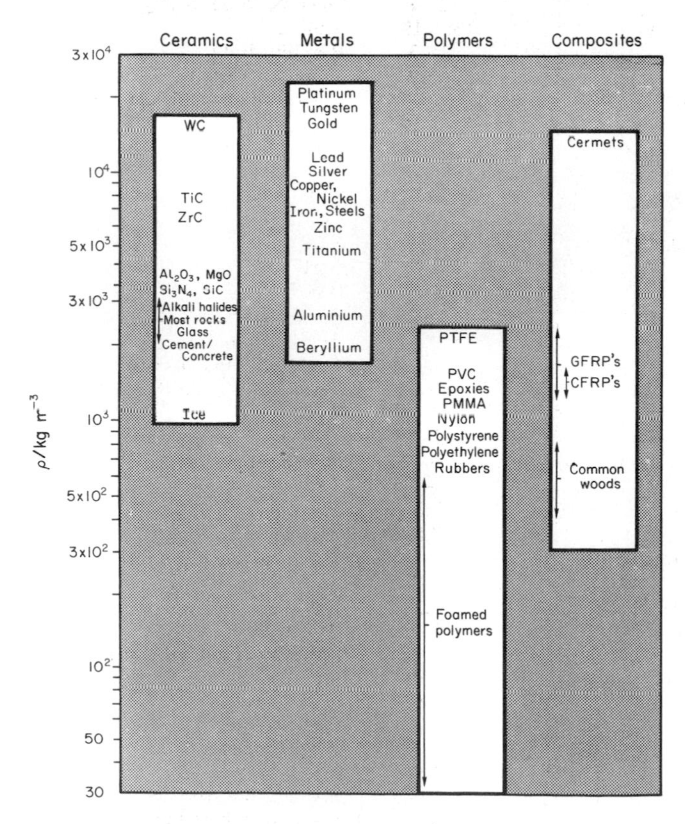

Fig. 5.12. Bar-chart of data for density, ρ.

Further reading

A. H. Cottrell, *Mechanical Properties of Matter*, Wiley, 1964, Chap. 3.
K. J. Pascoe, *An Introduction to the Properties of Engineering Materials*, 3rd edition, Van Nostrand, 1978, Chaps. 4, 20, 21.
C. Barrett and T. B. Massalski, *Structure of Metals*, 3rd edition, McGraw Hill, 1966 (for metals).
W. D. Kingery, *Introduction to Ceramics*, Wiley, 1960 (for ceramics).
I. M. Ward, *Mechanical Properties of Solid Polymers*, Wiley, 1971 (for polymers).

CHAPTER 6

THE PHYSICAL BASIS OF YOUNG'S MODULUS

Introduction

We are now in a position to bring together the factors underlying the moduli of materials. First, let us look back to Fig. 3.5, the bar-chart showing the moduli of materials. Recall that most ceramics and metals have moduli in a comparatively narrow range: 30–300 GN m^{-2}. Cement and concrete (45 GN m^{-2}) are near the bottom of that range. Aluminium (69 GN m^{-2}) is higher up; and steels (200 GN m^{-2}) are near the top. Special materials, it is true, lie outside it—diamond and tungsten lie above; ice and lead lie a little below—but most crystalline materials lie in that fairly narrow range. But polymers, all of them, lie below, some by several orders of magnitude. Why is this? What determines the general level of the moduli of solids? And is there the possibility of producing stiff polymers?

We shall now examine the modulus of ceramics, metals, polymers and composites, relating it to their structure.

Moduli of crystals

As we showed in Chapter 4, atoms in crystals are held together by electronic forces, or bonds, which resemble little springs. We defined the stiffness of one of these bonds as

$$S_0 = \left(\frac{\mathrm{d}^2 U}{\mathrm{d}r^2}\right)_{r=r_0}. \quad (6.1)$$

For *small* strains, S_0 stays constant (it is the *spring constant* of the bond). This means that the force between a pair of atoms, stretched apart to a distance r ($r \approx r_0$), is

$$F = S_0(r - r_0). \quad (6.2)$$

Imagine, now, a solid held together by such little springs, linking atoms across a plane within the material as shown in Fig. 6.1. To be correct, of course, we should draw out the atoms in the positions dictated by the *crystal structure* of a particular material. But for simplicity it is convenient just to put atoms at the corners of cubes of side r_0. In practice very few materials have such a simple crystal structure; but we shall not be *too* far out in our calculations by making our simplifying assumption—and it makes drawing the physical situation considerably easier!

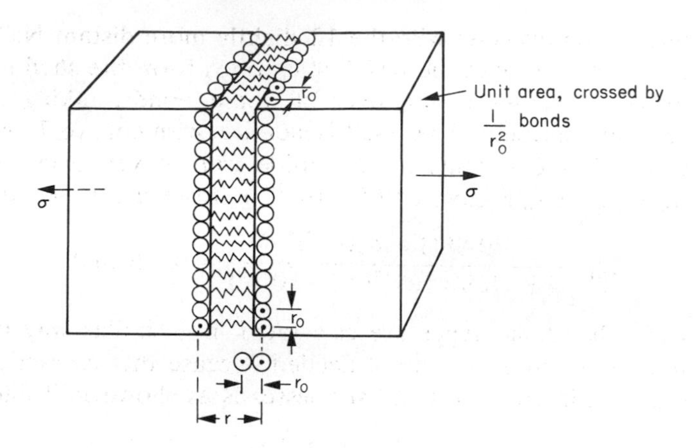

Fig. 6.1. Calculating Young's modulus from the stiffnesses of individual bonds.

Now, the total force exerted across *unit area*, if the two planes are pulled apart a distance $(r - r_0)$ is defined as the stress σ, with

$$\sigma = NS_0(r - r_0) \tag{6.3}$$

N is the number of bonds/unit area, equal to $1/r_0^2$ (since r_0^2 is the average area-per-atom). We convert displacement $(r - r_0)$ into strain ε_n by dividing by the *initial* spacing, r_0, so that

$$\sigma = \left(\frac{S_0}{r_0}\right)\varepsilon_n. \tag{6.4}$$

Young's modulus, then, is just

$$E = \frac{S_0}{r_0}. \tag{6.5}$$

S_0 can be calculated from the theoretically derived $U(r)$ curves. This is the realm of the solid-state physicist and quantum chemist, but we shall consider one example: the ionic bond, for which $U(r)$ is given in eqn. (4.3). Differentiating once with respect to r gives the *force* between the atoms, which must, of course, be zero at $r = r_0$. This gives us the constant B:

$$B = \frac{q^2 r_0^{n-1}}{4\pi n \varepsilon_0}. \tag{6.6}$$

Then eqn. (6.1) for S_0 gives

$$S_0 = \frac{\alpha q^2}{4\pi\varepsilon_0 r_0^3} \tag{6.7}$$

where $\alpha = (n - 1)$. But the coulombic attraction is a *long-range* interaction (it varies as $1/r$; an example of a short-range interaction is one which varies as $1/r^{10}$). Because of this, a given Na^+ ion not only interacts (attractively) with its shell of 6 neighbouring Cl^-

ions, it also interacts (repulsively) with the 12 slightly more distant Na^+ ions, with the 8 Cl^- ions beyond that, and with the 6 Na^+ ions which form the sheil beyond *that.* To calculate S_0 properly, we must sum over all these bonds, taking attractions and repulsions properly into account. The result is identical with eqn. (6.7), with $\alpha = 0.58$.

The Table of Physical Constants on the inside front cover gives value for q and ε_0; and r_0, the atom spacing, is close to 2.5×10^{-10} m. Inserting these values gives:

$$S_0 = \frac{0.58(1.6 \times 10^{-19})^2}{4\pi \times 8.8 \times 10^{-12}(2.5 \times 10^{-10})^3} = 9.5\ \mathrm{N\,m^{-1}}.$$

The stiffnesses of other bond types are calculated in a similar way (in general, the cumbersome sum described above is not needed because the interactions are of *short range*). The resulting hierarchy of bond stiffnesses is as shown in Table 6.1.

TABLE 6.1

Bond type	$S_0/\mathrm{N\,m^{-1}}$	Approximate E calculated from $(S_0/r_0)/\mathrm{GN\,m^{-2}}$
Covalent, C—C bond	180	1000
Pure ionic, e.g. Na—Cl bond	9–21	30–70
Pure metallic, e.g. Cu—Cu bond	15–40	30–150
H-bond, e.g. H_2O—H_2O	2	8
Van der Waals (waxes, many polymers)	1	2

Compare these predicted values of E with the measured values plotted in the bar-chart of Fig. 3.5. The comparison shows that, for metals and ceramics, the values of E we calculate are about right: the bond-stretching idea explains the stiffness of these solids. But a paradox remains: *there exists a whole range of polymers and rubbers which have moduli which are lower—by up to a factor of* 100*—than the lowest we have calculated.* Why is this? What determines the moduli of these floppy polymers if it is not the springs between the atoms? We shall explain this under our next heading.

Rubbers and the glass transition temperature

All polymers, if really solid, should have moduli above the lowest level we have calculated—about 2 $\mathrm{GN\,m^{-2}}$. If you take ordinary rubber tubing (a polymer) and cool it down in liquid nitrogen, it becomes stiff—its modulus rises rather suddenly from around $10^{-2}\ \mathrm{GN\,m^{-2}}$ to a "proper" value of about 4 $\mathrm{GN\,m^{-2}}$. But if you warm it up again, its modulus drops back to $10^{-2}\ \mathrm{GN\,m^{-2}}$.

This is because rubber, like many polymers, is composed of long spaghetti-like chains of carbon atoms, all tangled together as we showed in Chapter 5. In the case of rubber, the chains are also lightly cross-linked, as shown in Fig. 5.10. As we said in Chapter 5, along the carbon chain, and where there are occasional cross-links, are covalent bonds—very stiff bonds. But these contribute very little to the overall stiffness because when you load the structure it is the very much floppier Van der Waals bonds *between* the chains which stretch, and so the modulus is that of Van der Waals bonding, not of covalent bonding.

Well, that is the case at the low temperature, when the rubber has a "proper" modulus. As the rubber warms up to room temperature, the Van der Waals bonds *melt*. (In fact, the stiffness of the bond is proportional to its melting point: that is why diamond, which has the highest melting point of any material, also has the highest modulus.) The rubber remains solid because of the cross-links which form a sort of the skeleton: but when you load it, the chains can now slide over each other in places where there are no cross-linking bonds. This, of course, gives extra strain, and the modulus, which is σ/ε_n, goes down.

Many of the most floppy polymers have half-melted in this way at room temperature. The temperature at which this happens is called the *glass temperature*, T_G, for the polymer. Some polymers, which have no cross-links, melt completely at T_G, becoming viscous liquids. Others, containing cross-links, become *leathery* (like PVC) or rubbery (as polystyrene butadiene does). Some typical values for T_G are: polymethylmethacrylate (PMMA, or perspex), 100°C; polystyrene (PS), 90°C; polyethylene (low-density form), −20°C; *natural* rubber, −40°C. To summarise, above T_G, the polymer is leathery, rubbery or molten; below, it is a true solid with a modulus of at least $2\ \text{GN m}^{-2}$. This behaviour is shown in Fig. 6.2 which also shows how the stiffness of polymers increases as the covalent cross-link density increases, towards the value for diamond (which is simply a polymer with 100% of its bonds cross-linked, Fig. 4.7). Stiff polymers, then, *are* possible; the stiffest now available have moduli comparable with that of aluminium.

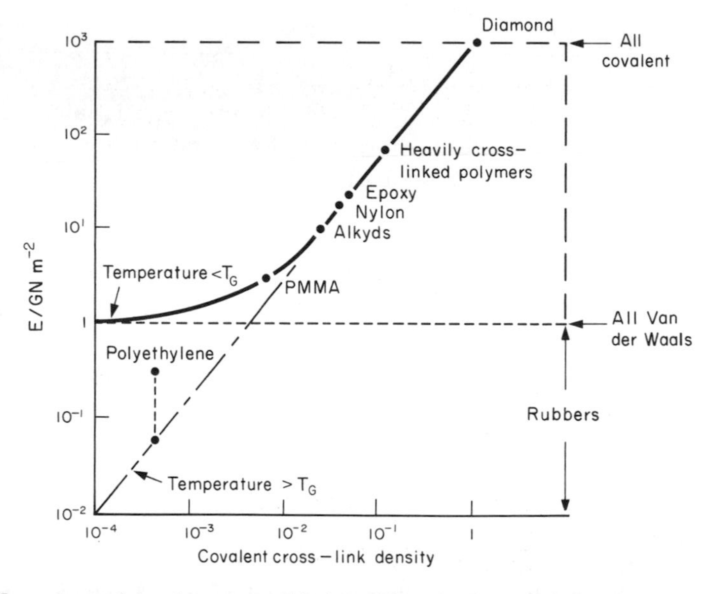

Fig. 6.2. How Young's modulus increases with increasing density of covalent cross-links in polymers, including rubbers above the glass temperature. Below T_G, the modulus of rubbers increases markedly due to the formation of Van der Waals cross-links with density approaching 1.

Composites

Is it possible to make polymers stiffer than the Van der Waals bonds which usually hold them together? The answer is yes—if we mix into the polymer a second, much stiffer, material. Good examples of materials stiffened in this way are:

(a) GFRP—glass-fibre-reinforced polymers, where the polymer is stiffened or reinforced by long fibres of soda glass;

(b) CFRP—carbon-fibre-reinforced polymers, where the reinforcement is achieved with fibres of graphite;

(c) BFRP—boron-fibre-reinforced polymers, using boron fibres as stiffening;

(d) FILLED POLYMERS—polymers into which glass powder or silica flour has been mixed to stiffen them;

(e) WOOD—a natural composite of lignin (an amorphous polymer) stiffened with fibres of cellulose.

The bar-chart of moduli (Fig. 3.5) shows that composites can have moduli much higher than those of their matrices. And it also shows that they can be *very* anisotropic: for example, the modulus of wood, measured parallel to the fibres, is about 10 GN m^{-2}; at right angles to this, it is less than 1 GN m^{-2}.

There is a very simple way to estimate the modulus of a fibre-reinforced composite. Suppose we stress a composite, containing a volume fraction V_f of fibres, parallel to the fibres (see Fig. 6.3(a)). It is natural to assume that the strains in the fibres and the matrix are equal. Then the stress carried by the composite is

$$\sigma = V_f\sigma_f + (1 - V_f)\sigma_m$$
$$= E_f V_f \varepsilon_n + E_m(1 - V_f)\varepsilon_n.$$

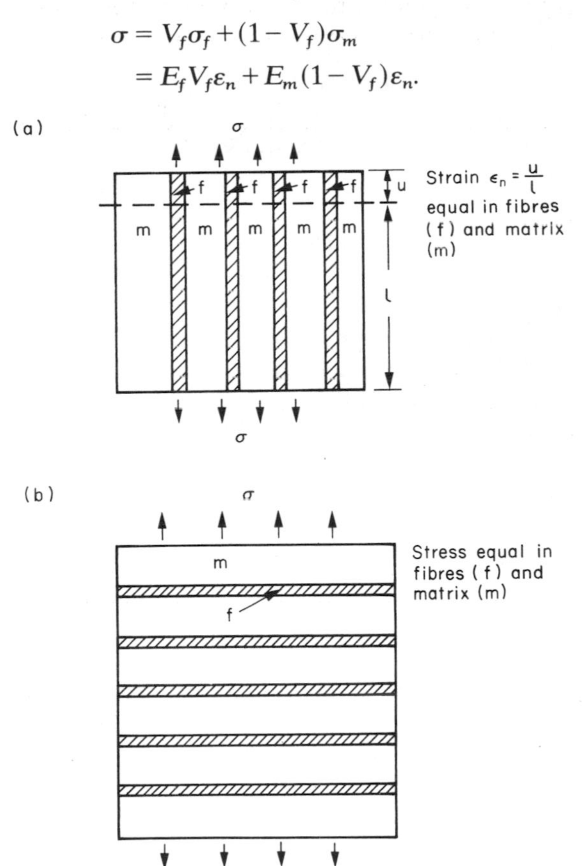

Fig. 6.3. A fibre-reinforced composite loaded so as to give (a) maximum modulus, (b) minimum modulus.

But since $E_{\text{composite}} = \sigma/\varepsilon_n$, we find

$$E_{\text{composite}} = V_f E_f + (1 - V_f) E_m. \tag{6.8}$$

Obviously this gives us an upper estimate for the modulus of our fibre-reinforced composite. The modulus cannot be greater than this value.

How is it that the modulus can be less? Suppose we had loaded the composite (Fig. 6.3(b)) in the opposite way, at right angles to the fibres? It now becomes much more reasonable to assume that the *stresses*, not the strains, in the two components are equal. If this is so, then the total nominal strain ε_n is the weighted sum of the individual strains:

$$\begin{aligned} \varepsilon_n &= V_f \varepsilon_{nf} + (1 - V_f) \varepsilon_{nm} \\ &= \frac{V_f \sigma}{E_f} + \left(\frac{1 - V_f}{E_m}\right) \sigma. \end{aligned}$$

The modulus, as before, is σ/ε_n so that

$$E_{\text{composite}} = 1 \Big/ \left\{ \frac{V_f}{E_f} + \frac{(1 - V_f)}{E_m} \right\}. \tag{6.9}$$

Although it is not obvious, this is a lower limit for the modulus—it cannot be less than this.

The two estimates, if plotted, look as shown in Fig. 6.4. This explains why fibre-reinforced composites like wood and GFRP are so stiff along the reinforced direction, and yet so floppy at right angles to the direction of reinforcement, i.e. why they are so *anisotropic*.

To conclude, we estimate the moduli of composites in which the matrix is stiffened by (roughly spherical) *particles* rather than by continuous *fibres*. The theory is, as one might imagine, more difficult than for fibre-reinforced composites; and is too advanced to talk about here. But it turns out that the moduli of these so-called *particulate* composites lie quite close to the lower value for fibre-reinforced composites, as shown in Fig. 6.4. Now, it is much cheaper to mix sand into a polymer than to carefully align specially produced glass fibres in the same polymer. Thus the modest

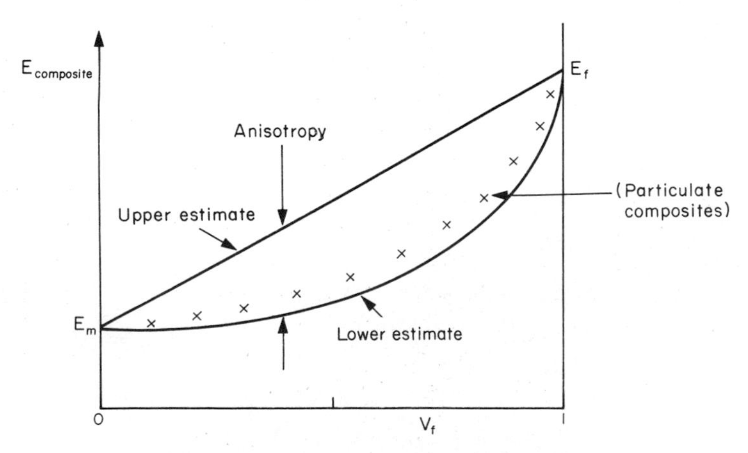

Fig. 6.4. Composite modulus for various volume fractions of stiffener.

increase in stiffness that adding particles gives is economically worth while. Naturally the resulting particulate composite is *isotropic*, rather than *anisotropic* as would be the case for the fibre-reinforced composites; and this, too, can be an advantage. These filled polymers can be formed and moulded by normal methods (most fibre-composites cannot) and so are cheap to fabricate. They are appearing in very recent models of cars as bumpers, grilles, and protective trim (filled polypropylene is the commonest).

Summary

The moduli of metals, ceramics and glassy polymers below T_G reflect the stiffness of the bonds which link the atoms. Glasses and glassy polymers above T_G are leathers, rubbers or viscous liquids, and have much lower moduli. Composites have moduli which are a weighted average of those of their components.

Further reading

A. H. Cottrell, *The Mechanical Properties of Matter*, Wiley, 1964, Chap. 4.
C. Kittel, *Introduction to Solid State Physics*, 4th edition, Wiley, 1971, Chaps. 3 and 4 (for metals and ceramics).
I. M. Ward, *Mechanical Properties of Solid Polymers*, Wiley, 1971, Chap. 3 (for polymers).

CHAPTER 7

CASE STUDIES OF MODULUS-LIMITED DESIGN

CASE STUDY 1: A TELESCOPE MIRROR—INVOLVING THE SELECTION OF A MATERIAL TO MINIMISE THE DEFLECTION OF A DISC UNDER ITS OWN WEIGHT

Introduction

The world's largest telescope is sited on Mount Semivodrike, near Zelenchukskaya in the Caucasus Mountains of the USSR. The mirror is 6 m in diameter (236 inches, 36 inches larger than the Western World's largest telescope at Mount Palomar). To be sufficiently rigid, the mirror (which is made of glass) is about 1 m thick and weighs 70 tonnes.

The cost of a large (200-inch) telescope is, like the telescope itself, astronomical—about UK£80 m or US$176 m. This cost varies roughly with the square of the weight of the mirror which, by itself, accounts for about 5% of the total cost of the telescope. The mechanism which holds, positions and moves the mirror as it tracks across the sky (Fig. 7.1) must be stiff enough that it can position the mirror relative to the collecting system with a precision about equal to that of the wavelength of light. Although, at first sight, if you double the mass M of the mirror, you need only double the sections of the structure which holds it in order to keep the stresses (and hence the strains) the same, the heavier structure then deflects under its own weight. In practice, you have to make linear dimensions of the structure increase with the mass of the mirror so that the volume (and thus the cost) goes as M^2. The main obstacle to building such large telescopes is the cost.

Before the turn of the century, mirrors were made of speculum metal. Since then, they have been made of glass, silvered on the *front* surface, so none of the optical properties of the glass are used. Glass is chosen for its mechanical properties only; the 70 tonnes of glass is just a very elaborate support for 100 nm (about 30 g) of silver. Could one, by taking a radically new look at mirror design, suggest possible routes to the construction of larger mirrors which are much lighter than the present ones?

Optimum combination of elastic properties for the mirror support

Consider the selection of the material for the mirror backing of a 200-inch (5 m) diameter telescope. We want to identify the material that gives a mirror which will

Fig. 7.1. The new British infra-red telescope at Mauna Kea, Hawaii. The picture shows the housing for the 3.8 m diameter mirror, the supporting frame, and the interior of the aluminium dome with its sliding "window". (© 1979 by Photolabs, Royal Observatory, Edinburgh.)

suffer minimal distortion when it is moved, and has minimum weight. We will limit ourselves to these criteria alone for the moment—we will leave the problem of grinding the parabolic shape and getting an optically perfect surface to the development research team.

At its simplest, the mirror is a circular disc, of diameter $2a$ and mean thickness t, simply supported at its periphery. When horizontal, it will deflect under its own weight M; when vertical it will not deflect significantly. We want this distortion (which changes the focal length and introduces aberrations into the mirror) to be small enough that it does not significantly reduce the performance of the mirror. In practice, this means that the deflection δ of the mid-point of the mirror must be less than the wavelength of light. We shall require, therefore, that the mirror deflect less than $\approx 1\ \mu m$ at its centre. This is an exceedingly stringent limitation (Fig. 7.2). Fortunately, it can be *partially* overcome by engineering design without reference to the material used. By using counterbalanced weights or hydraulic jacks, the mirror can be supported by distributed

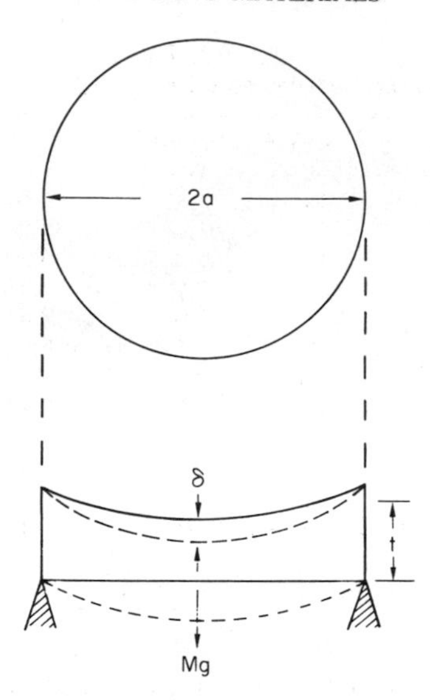

Fig. 7.2. Elastic deflection of a telescope mirror under its own weight.

forces over its back surface which are made to vary automatically according to the attitude of the mirror (Fig. 7.3). Nevertheless, the limitations of this compensating system still require that the mirror have a stiffness such that δ (if supported at its periphery) be less than 10 μm.

The formulae for the elastic deflections of plates and beams under their own weight you will find in standard texts on mechanics or structures. We need only one formula here: it is that the deflection, δ, of the centre of a horizontal disc due to its own weight is given quite well by

$$\delta = \frac{3}{4\pi} \frac{Mga^2}{Et^3}, \tag{7.1}$$

for a material having a Poisson's ratio fairly close to 0.33. The quantity g in this equation is the acceleration due to gravity. We need to minimise the (variable) mass for fixed values of $2a$ (5 m) and δ (10 μm). The thickness t is an added variable, but it can be expressed in terms of the mass by $M = \pi a^2 t \rho$, or

$$t = \frac{M}{\pi a^2 \rho}, \tag{7.2}$$

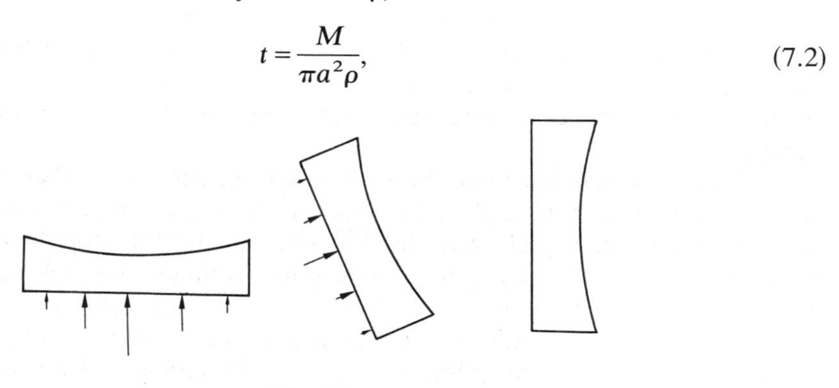

Fig. 7.3.

TABLE 7.1
MIRROR BACKING FOR 200-INCH TELESCOPE

Material	E/GN m^{-2}	ρ/Mg m^{-3}	$(\rho^3/E)^{1/2}$/Ns^3 m^{-5}	M/tonne	t/m
Steel	200	7.8	1.54	158	1.0
Concrete	47	2.5	0.56	56	1.2
Aluminium	69	2.7	0.53	53	1.0
Glass	69	2.5	0.48	48	0.97
GFRP	40	2.0	0.45	44	1.1
Wood	12	0.6	0.13	14	1.2
Foamed polyurethane	0.06	0.1	0.13	13	6.6
CFRP	270	1.5	0.11	11	0.38

where ρ is the material density. If we substitute for t in eq. (7.1), we have

$$\delta = \frac{3}{4\pi}\frac{Mga^2}{E}\frac{\pi^3 a^6 \rho^3}{M^3}. \tag{7.3}$$

Finally, the mass can be extracted from this equation to give

$$M = \left(\frac{3g}{4\delta}\right)^{1/2} \pi a^4 \left(\frac{\rho^3}{E}\right)^{1/2}. \tag{7.4}$$

Clearly, the only variables left on the right-hand side of eqn. (7.4) are the *material* properties ρ and E. To minimise M, we must choose a material having the minimum possible value of $(\rho^3/E)^{1/2}$.

Let us now examine values of $(\rho^3/E)^{1/2}$ for some materials. Data for E we can take from Table 3.1 in Chapter 3; those for density, from Table 5.1 in Chapter 5. The results are as shown in Table 7.1.

Conclusion

The optimum material is CFRP. The next best is polyurethane foam. Glass is better than steel, aluminium or concrete (that is why most mirrors are made of glass), but not as good as wood.

We should, of course, examine other aspects of this choice. The mass of the mirror can be calculated from eqn. (7.4) for the various materials listed in Table 7.1. Note that the polyurethane foam and the CFRP mirrors are roughly one-fifth the weight of the glass one, and that the structure needed to support a CFRP mirror could thus be as much as 25 times less expensive than the structure needed to support an orthodox glass mirror.

In addition, we can calculate the thickness from eqn. (7.2). Values of t for various materials are given in Table 7.1. The glass mirror has to be about 1 m thick (and real mirrors are about this thick); the CFRP-backed mirror need only be 0.38 m thick. The polyurethane foam mirror has to be very thick—although there is no reason why one could not make a 6 m cube of such a foam.

Some of the above solutions—such as the use of polyurethane foam for mirrors—may at first seem ridiculously impractical. But the potential cost-saving UK£3.2 m or US$7 m per telescope in place of UK£80 m or US$176 m is so vast that they are worth examining closely. There are ways of casting a thin film of silicone rubber, or of epoxy, onto the surface of the mirror-backing (the polyurethane or the CFRP) to give an optically smooth surface which could be silvered. The most obvious obstacle is the lack of stability of polymers—they change dimensions with age, humidity, temperature and so on. But glass itself can be foamed to give a material with a density not much larger than polyurethane foam, and the same stability as solid glass, so a study of this sort can suggest radically new solutions to design problems by showing how new classes of materials might be used.

Case study 2: materials selection to give a beam of a given stiffness with minimum weight

Introduction

Many structures require that a beam sustain a certain force F without deflecting more than a given amount, δ. If, in addition, the beam forms part of a transport system—a plane or rocket, or a train—or something which has to be carried or moved—a rucksack for instance—then it is desirable, also, to minimise the weight.

In the following, we shall consider a single cantilever beam, of square section, and will analyse the material requirements to minimise the weight for a given stiffness. The results are quite general in that they apply equally to any sort of beam of square section, and can easily be modified to deal with beams of other sections: tubes, I-beams, box-sections and so on.

Analysis

The square-section beam of length l (determined by the design of the structure, and thus fixed) and thickness t (a variable) is held rigidly at one end while a force F (the maximum service force) is applied to the other, as shown. The (elastic) deflection is

$$\delta = \frac{4l^3F}{Et^4}, \tag{7.5}$$

ignoring self-weight.

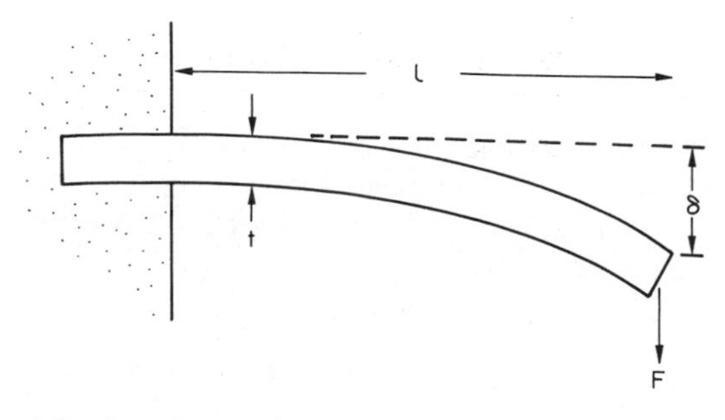

Fig. 7.4. Elastic deflection of a cantilever beam under an externally imposed force F.

The mass is given by

$$M = lt^2\rho \tag{7.6}$$

so that

$$t = \left(\frac{M}{l\rho}\right)^{1/2}. \tag{7.7}$$

Substituting for t in eqn. (7.5) we have

$$\delta = \frac{4l^3F}{E}\frac{l^2\rho^2}{M^2} \tag{7.8}$$

which gives for the mass

$$M = \left(\frac{4l^5F}{\delta}\right)^{1/2}\left(\frac{\rho^2}{E}\right)^{1/2}. \tag{7.9}$$

The mass of the beam, for given stiffness F/δ, is minimised by selecting a material with the minimum $(\rho^2/E)^{1/2}$.

Conclusions

Table 7.2 gives values for $(\rho^2/E)^{1/2}$. Among cheap materials, wood is one of the best bets—that is why it is so widely used in small-scale building, for the handles of rackets and shafts of golf-clubs, for vaulting poles, even for building aircraft. Polyurethane foam is no good at all—the criteria here are quite different than in the first case study. The only material which is clearly superior to wood is CFRP—and it would reduce the mass of the beam very substantially: by the factor 5.5/2.9, or very nearly a factor of 2. That is why CFRP is used when weight-saving is the overriding design criterion. But as we shall see in a moment, it is very expensive.

Why, then, are bicycles not made of wood? (There was a time when they were.) That is because metals, and polymers, too, can readily be made in tubes. The formula for the bending of a tube depends on the mass of the tube in a different way than does that of a solid beam, and the optimisation we have just performed—which is easy enough to do—favours the tube.

TABLE 7.2
DATA FOR BEAM OF GIVEN STIFFNESS

Material	$(\rho^2/E)^{1/2} \times 10^3$/$N^{1/2}\,m^{-3}\,s^2$	$\tilde{p}$/UK£ tonne^{-1} (US\$ tonne^{-1})	$\tilde{p}(\rho^2/E)^{1/2} \times 10^3$/ UK£ (US\$) $N^{-1/2}\,m^{-2}$
Concrete	12	130 (290)	1.6 (3.5)
Wood	5.5	196 (431)	1.1 (2.4)
Steel	17	206 (453)	3.5 (7.7)
Aluminium	10	1,060 (2330)	11 (24)
GFRP	10	1,500 (3300)	15 (33)
CFRP	2.9	90,000 (198,000)	261 (574)
Polyurethane foam	13	500 (1100)	6.5 (14)

Case study 3: Materials selection to minimise cost of a beam of given stiffness

Introduction

Often it is not the weight, but the *cost* of a structure which is the overriding criterion. Suppose that had been the case with the cantilever beam that we have just considered—would our conclusion have been the same? Would we still select wood? And how much more expensive would a replacement by CFRP be?

Analysis

The price per tonne, $\tilde{p}$, of the material of the beam is the first of the properties of materials that we talked about in our book. The total price of the beam, crudely, is the weight of the beam times $\tilde{p}$ (though this may neglect certain aspects of manufacture). Thus

$$\text{Price} = \left(\frac{4l^5F}{\delta}\right)^{1/2} \tilde{p}\left(\frac{\rho^2}{E}\right)^{1/2}. \tag{7.10}$$

The beam of minimum price is therefore the one with the lowest value of $\tilde{p}(\rho^2/E)^{1/2}$. Values for this are given in Table 7.2, with prices taken from the table in Chapter 2.

Conclusions

Concrete and wood are the cheapest materials to use for a beam having a given stiffness. Steel costs more; but it can be rolled to give I-section beams which have a much better stiffness-to-weight ratio than the solid square-section beam we have been analysing here. This compensates for steel's rather high cost, and accounts for the interchangeable use of steel, wood and concrete that we talked about in bridge construction in Chapter 1. Finally, the lightest beam (CFRP) costs more than 200 times that of a wooden one—and this cost at present rules out CFRP for all but the most specialised applications like aircraft components or ultra-sophisticated sporting equipment. But the cost of CFRP falls as the market for it expands. If (as now seems possible) it is incorporated into automobiles (to lighten them and so save fuel), its price will fall to a level at which it will compete with metals in many applications.

Further reading

A. H. Cottrell, *Mechanical Properties of Matter*, Wiley, 1964, Chap. 5.
S. P. Timoshenko and J. N. Goodier, *Theory of Elasticity*, 3rd edition, McGraw Hill, 1970.

C. Yield strength, tensile strength, hardness and ductility

CHAPTER 8

THE YIELD STRENGTH, TENSILE STRENGTH, HARDNESS AND DUCTILITY

Introduction

All solids have an *elastic limit* beyond which something happens. A totally brittle solid will fracture, either suddenly (like glass) or progressively (like cement or concrete). Most engineering materials do something different; they deform *plastically* or change their shapes in a *permanent* way. It is important to know when, and how, they do this—both so that we can design structures which will withstand normal service loads without any permanent deformation, and so that we can design rolling mills, sheet presses, and forging machinery which will be strong enough to impose the desired deformation onto materials we wish to form. To study this, we pull carefully prepared samples in a tensile-testing machine, or compress them in a compression machine (which we will describe in a moment), and record the *stress* required to produce a given *strain.*

Linear and non-linear elasticity; anelastic behaviour

Figure 8.1 shows the *stress–strain* curve of a material exhibiting *perfectly linear elastic* behaviour. This is the behaviour characterized by Hooke's Law (Chapter 3). All solids

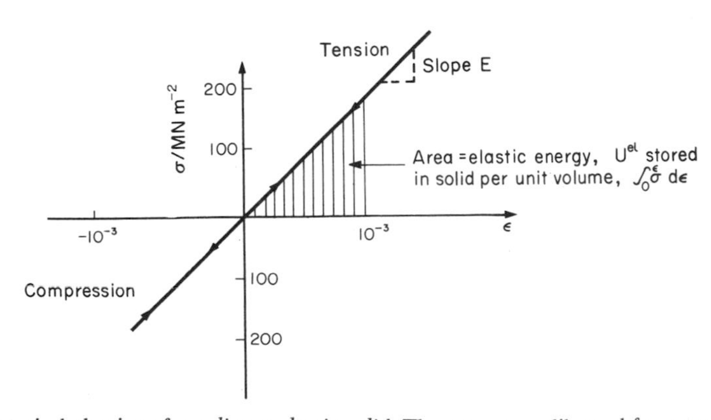

Fig. 8.1. Stress–strain behaviour for a *linear elastic solid.* The axes are calibrated for a material such as steel.

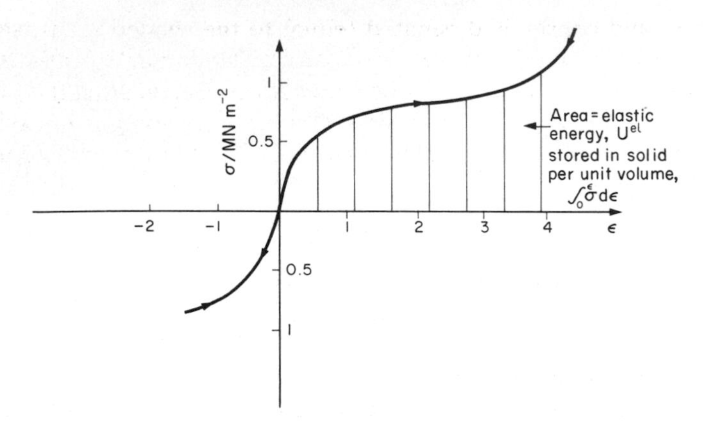

Fig. 8.2. Stress–strain behaviour for a *non-linear elastic solid.* The axes are calibrated for a material such as rubber.

are linear elastic at small strains—by which we usually mean less than 0.001, or 0.1%. The slope of the stress–strain line, which is the same in compression as in tension, is of course Young's Modulus, *E*. The area (shaded) is the elastic energy stored, per unit volume: since it is an elastic solid, we can get it all back if we unload the solid, which behaves like a linear spring.

Figure 8.2 shows a *non-linear* elastic solid. *Rubbers* have a stress–strain curve like this, extending to very large strains (of order 5). The material is still elastic: if unloaded, it follows the same path down as it did up, and all the energy stored, per unit volume, during loading is recovered on unloading—that is why catapults can be as lethal as they are.

Finally, Fig. 8.3 shows a third form of *elastic* behaviour found in certain materials. This is called *anelastic* behaviour. All solids are anelastic to a small extent: even in the régime where they are nominally elastic, the loading curve does not *exactly* follow the

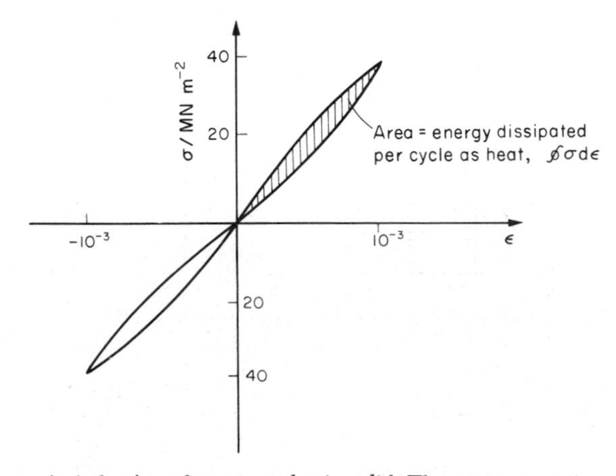

Fig. 8.3. Stress–strain behaviour for an *anelastic solid.* The axes are calibrated for fibreglass.

unloading curve, and energy is dissipated (equal to the shaded area) when the solid is cycled. Sometimes this is useful—if you wish to damp out vibrations or noise, for example; you can do so with polymers or with soft metals (like lead) which have a high *damping capacity* (high anelastic loss). But often such damping is undesirable—springs and bells, for instance, are made of materials with the lowest possible damping capacity (spring steel, bronze, glass).

Load–extension curves for non-elastic (plastic) behaviour

Rubbers are exceptional in behaving reversibly, or *almost* reversibly, to high strains; as we said, *almost all materials, when strained by more than about* 0.001 (0.1%), *do something irreversible*: and most engineering materials deform *plastically* to change their shape *permanently*. If we load a piece of ductile metal (like copper), for example in tension, we get the following relationship between the load and the extension (Fig. 8.4). This can be demonstrated nicely by pulling a piece of plasticine (a ductile non-metallic material). Initially, the plasticine deforms elastically, but at a small strain begins to deform plastically, so that if the load is removed, the piece of plasticine is permanently longer than it was at the beginning of the test: it has undergone *plastic* deformation (Fig. 8.5). If you continue to pull, it continues to get longer, at the same time getting thinner because in plastic deformation *volume is conserved* (matter is just flowing from place to place). Eventually, the plasticine becomes unstable and begins to *neck* at the maximum load point in the force–extension curve (Fig. 8.4). Necking is an *instability* which we shall look at in more detail in Chapter 11. The neck then grows quite rapidly, and the load that the specimen can bear through the neck decreases until breakage takes place. The two pieces produced *after* breakage have a total length that is slightly *less* than the length *just* before breakage by the amount of the *elastic* extension produced by the terminal load.

If we load a material in *compression*, the force–displacement curve is simply the reverse of that for tension *at small strains*, but it becomes different at larger strains.

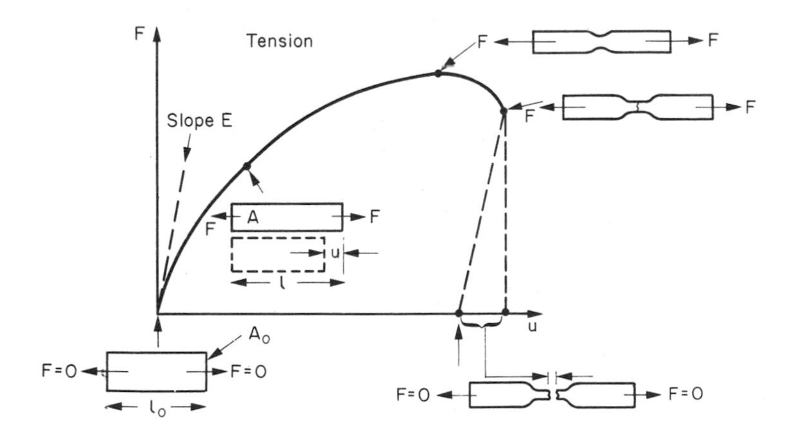

Fig. 8.4. Load–extension curve for a bar of ductile metal (e.g. annealed copper) pulled in tension.

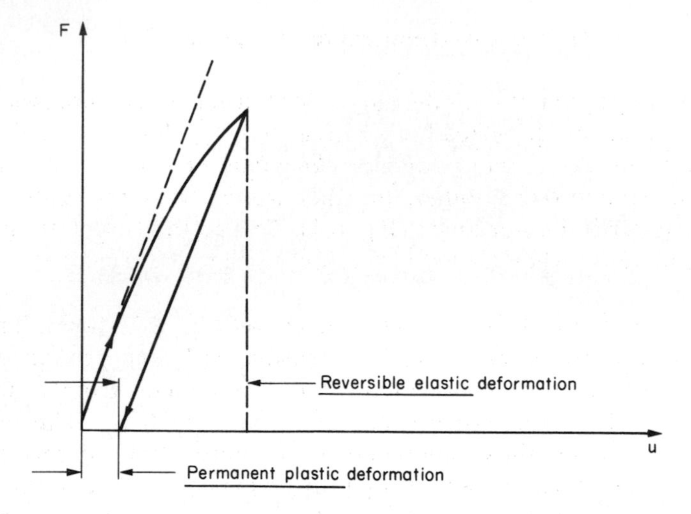

Fig. 8.5.

As the specimen squashes down, becoming shorter and fatter to conserve volume, the load needed to keep it flowing rises (Fig. 8.6). No instability such as necking appears, and the specimen can be squashed almost indefinitely, this process only being limited eventually by severe cracking in the specimen or the plastic flow of the compression plates.

Why this great difference in behaviour? After all, we are dealing with the same material in either case.

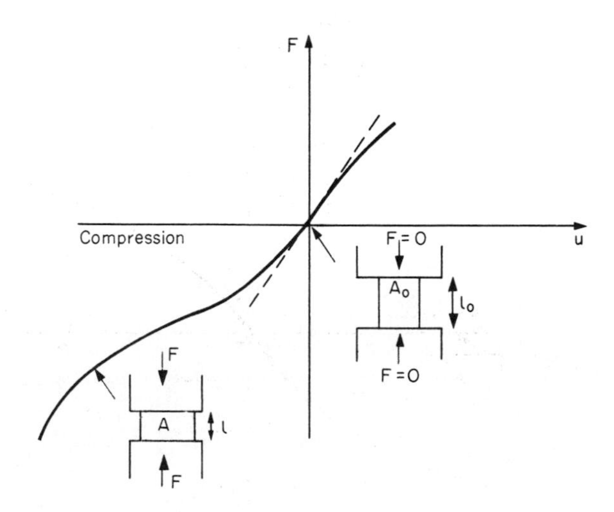

Fig. 8.6.

True stress–strain curves for plastic flow

The apparent difference between the curves for tension and compression is due solely to the geometry of testing. If, instead of plotting *load*, we plot *load divided by the actual area of the specimen, A, at any particular elongation or compression*, the two curves become much more like one another. In other words, we simply plot *true stress* (see Chapter 3) as our vertical co-ordinate (Fig. 8.7). This method of plotting allows for the *thinning* of the material when pulled in tension, or the *fattening* of the material when compressed.

But the two curves still do not exactly match, as Fig. 8.7 shows. The reason is a displacement of (for example) $u = l_0/2$ in tension and compression gives different *strains*; it represents a drawing out of the tensile specimen from l_0 to $1.5l_0$, but a squashing down of the compressive specimen from l_0 to $0.5l_0$. The material of the compressive specimen has thus undergone *much* more plastic deformation than the material in the tensile specimen, and can hardly be expected to be in the same state, or to show the same resistance to plastic deformation. The two conditions can be compared properly by taking small *strain increments*

$$\delta\varepsilon = \frac{\delta u}{l} = \frac{\delta l}{l} \tag{8.1}$$

about which the state of the material is the same for either tension or compression (Fig. 8.8). This is the same as saying that a decrease in length from 100 mm (l_0) to 99 mm (l), or an increase in length from 100 mm (l_0) to 101 mm (l) both represent a 1% change in the state of the material. Actually, they do not *quite* give exactly 1% in both cases, of course, but they *do* in the limit

$$\mathrm{d}\varepsilon = \frac{\mathrm{d}l}{l}. \tag{8.2}$$

Then, if the stresses in compression and tension are plotted against

$$\varepsilon = \int_{l_0}^{l} \frac{\mathrm{d}l}{l} = \ln\left(\frac{l}{l_0}\right) \tag{8.3}$$

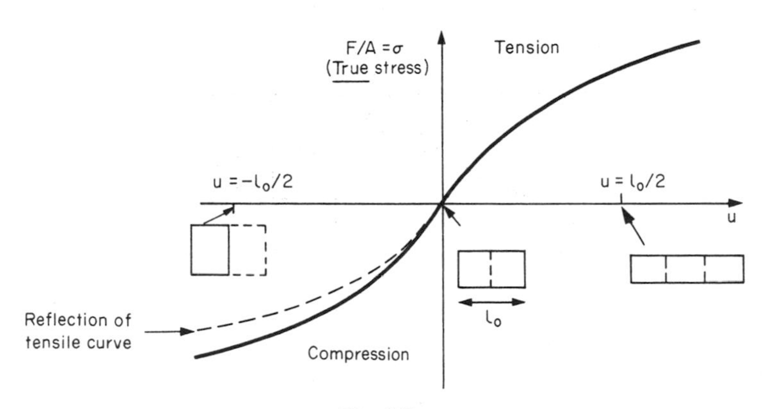

Fig. 8.7.

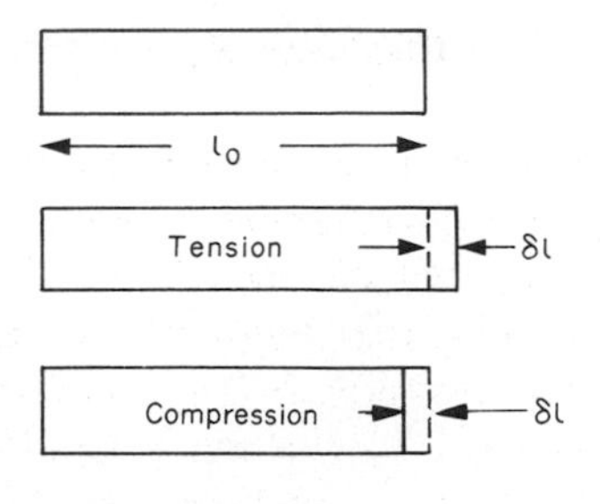

Fig. 8.8.

the two curves *exactly* mirror one another (Fig. 8.9). The quantity ε is called the *true* strain (to be contrasted with the *nominal* strain u/l_0 (defined in Chapter 3)) and the matching curves are *true stress/true strain* (σ/ε) curves. Now, a final catch. We can, from our original load–*extension* or load–*compression* curves easily calculate ε, simply by knowing l_0 and taking natural logs. But how do we calculate σ? Because volume is conserved during plastic deformation we can write, at any strain,

$$A_0 l_0 = Al$$

provided the extent of plastic deformation is much greater than the extent of elastic deformation (this is usually the case, but the qualification must be mentioned because volume is only conserved during *elastic* deformation if Poisson's ratio $\nu = 0.5$; and, as we showed in Chapter 3, it is near 0.33 for most materials). Thus

$$A = \frac{A_0 l_0}{l} \tag{8.4}$$

and

$$\sigma = \frac{F}{A} = \frac{Fl}{A_0 l_0}, \tag{8.5}$$

all of which we know or can measure easily.

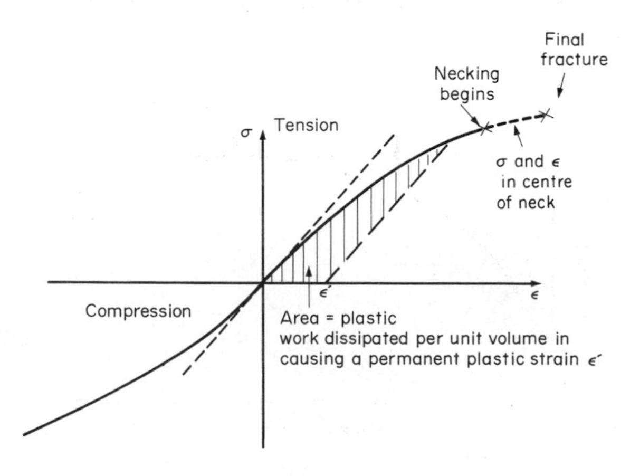

Fig. 8.9.

Plastic work

When metals are rolled or forged, or drawn to wire, or when polymers are injection-moulded or pressed or drawn, energy is absorbed. The work done on a material to change its shape permanently is called the *plastic work*; its value, per unit volume, is the area of the cross-hatched region shown in Fig. 8.9; it may easily be found (if the stress–strain curve is known) for any amount of permanent plastic deformation, ε'. Plastic work is important in metal- and polymer-forming operations because it determines the forces that the rolls, or press, or moulding machine must exert on the material.

Tensile testing

The plastic behaviour of a material is usually measured by conducting a tensile test. Tensile testing equipment is standard in all engineering laboratories. Such equipment produces a load/displacement (F/u) curve for the material, which is then converted to a nominal stress/nominal strain, or σ_n/ε_n, curve (Fig. 8.10), where

$$\sigma_n = \frac{F}{A_0} \tag{8.6}$$

and

$$\varepsilon_n = \frac{u}{l_0} \tag{8.7}$$

(see Chapter 3, and above). Naturally, because A_0 and l_0 are constant, the *shape* of the σ_n/ε_n curve is identical to that of the load–extension curve. But the σ_n/ε_n plotting method allows one to compare data for specimens having different (though now standardised) A_0 and l_0, and thus to examine the properties of *material*, unaffected by specimen size. The advantage of keeping the stress in *nominal* units and not converting to *true* stress (as shown above) is that the onset of necking can clearly be seen on the σ_n/ε_n curve.

Now, let us define the quantities usually listed as the results of a *tensile test*. The easiest way to do this is to show them on the σ_n/ε_n curve itself (Fig. 8.11). They are:

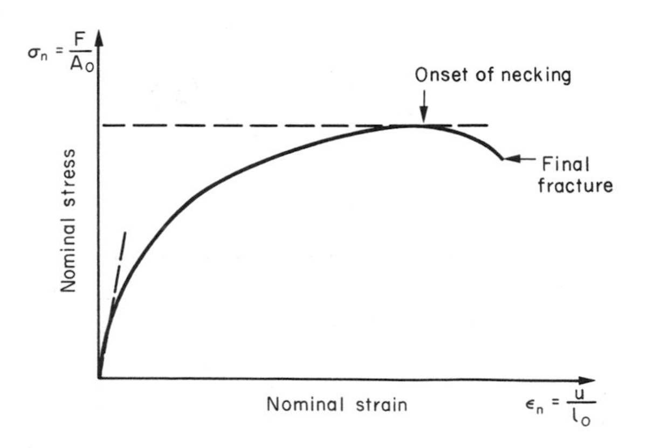

Fig. 8.10.

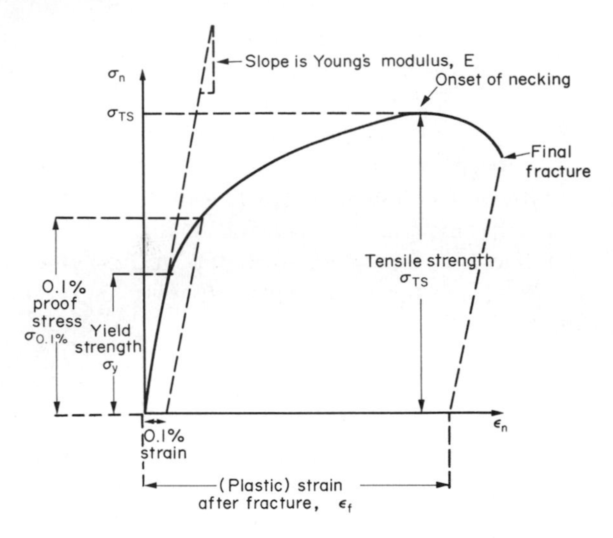

Fig. 8.11.

σ_y *Yield strength* (F/A_0 at onset of plastic flow).

$\sigma_{0.1\%}$ *0.1% Proof stress* (F/A_0 at a permanent strain of 0.1%) (0.2% proof stress is often quoted instead. Proof stress is useful for characterising yield of a material that yields gradually, and does not show a distinct yield point.)

σ_{TS} *Tensile strength* (F/A_0 at onset of necking).

ε_f (Plastic) *strain after fracture*, or tensile ductility. The broken pieces are put together and measured, and ε_f calculated from $(l - l_0)/l_0$, where l is the length of the assembled pieces.

Data

Data for the *yield strength*, *tensile strength* and the *tensile ductility* are given in Table 8.1 and shown on the bar-chart (Fig. 8.12). Like moduli, they span a range of about

TABLE 8.1

YIELD STRENGTH, σ_y, TENSILE STRENGTH, σ_{TS}, AND TENSILE DUCTILITY, ε_f

Material	σ_y/MN m^{-2}	σ_{TS}/MN m^{-2}	ε_f
Diamond	50,000	—	0
Silicon carbide, SiC	10,000	—	0
Silicon nitride, Si_3N_4	8000	—	0
Silica glass, SiO_2	7200	—	0
Tungsten carbide, WC	6000	—	0
Niobium carbide, NbC	6000	—	0
Alumina, Al_2O_3	5000	—	0
Beryllia, BeO	4000	—	0
Mullite	4000	—	0
Titanium carbide, TiC	4000	—	0
Zirconium carbide, ZrC	4000	—	0
Tantalum carbide, TaC	4000	—	0
Zirconia, ZrO_2	4000	—	0
Soda glass (standard)	3600	—	0
Magnesia, MgO	3000	—	0
Cobalt and alloys	180–2000	500–2500	0.01–6

Table 8.1 (*Continued*)

Material	σ_y/MN m^{-2}	σ_{TS}/MN m^{-2}	ε_f
Low-alloy steels (water-quenched and tempered)	500–1980	680–2400	0.02–0.3
Pressure-vessel steels	1500–1900	1500–2000	0.3–0.6
Stainless steels, austenitic	286–500	760–1280	0.45–0.65
Boron/epoxy composites (tension–compression)	—	725–1730	—
Nickel alloys	200–1600	400–2000	0.01–0.6
Nickel	70	400	0.65
Tungsten	1000	1510	0.01–0.6
Molybdenum and alloys	560–1450	665–1650	0.01–0.36
Titanium and alloys	180–1320	300–1400	0.06–0.3
Carbon steels (water-quenched and tempered)	260–1300	500–1880	0.2–0.3
Tantalum and alloys	330–1090	400–1100	0.01–0.4
Cast irons	220–1030	400–1200	0–0.18
Copper alloys	60–960	250–1000	0.01–0.55
Copper	60	400	0.55
Cobalt/tungsten carbide cermets	400–900	900	0.02
CFRPs (tension–compression)	—	670–640	—
Brasses and bronzes	70–640	230–890	0.01–0.7
Aluminium alloys	100–627	300–700	0.05–0.3
Aluminium	40	200	0.5
Stainless steels, ferritic	240–400	500–800	0.15–0.25
Zinc alloys	160–421	200–500	0.1–1.0
Concrete, steel reinforced (tension or compression)	—	410	0.02
Alkali halides	200–350	—	0
Zirconium and alloys	100–365	240–440	0.24–0.37
Mild steel	220	430	0.18–0.25
Iron	50	200	0.3
Magnesium alloys	80–300	125–380	0.06–0.20
GFRPs	—	100–300	—
Beryllium and alloys	34–276	380–620	0.02–0.10
Gold	40	220	0.5
PMMA	60–110	110	—
Epoxies	30–100	30–120	—
Polyimides	52–90	—	—
Nylons	49–87	100	—
Ice	85	—	0
Pure ductile metals	20–80	200–400	0.5–1.5
Polystyrene	34–70	40–70	—
Silver	55	300	0.6
ABS/polycarbonate	55	60	—
Common woods (compression, ∥ to grain)	—	35–55	—
Lead and alloys	11–55	14–70	0.2–0.8
Acrylic/PVC	45–48	—	—
Tin and alloys	7–45	14–60	0.3–0.7
Polypropylene	19–36	33–36	—
Polyurethane	26–31	58	—
Polyethylene, high density	20–30	37	—
Concrete, non-reinforced, compression	20–30	—	0
Natural rubber	—	30	5.0
Polyethylene, low density	6–20	20	—
Common woods (compression, ⊥ to grain)	—	4–10	—
Ultrapure f.c.c. metals	1–10	200–400	1–2
Foamed polymers, rigid	0.2–10	0.2–10	0.1–1
Polyurethane foam	1	1	0.1–1

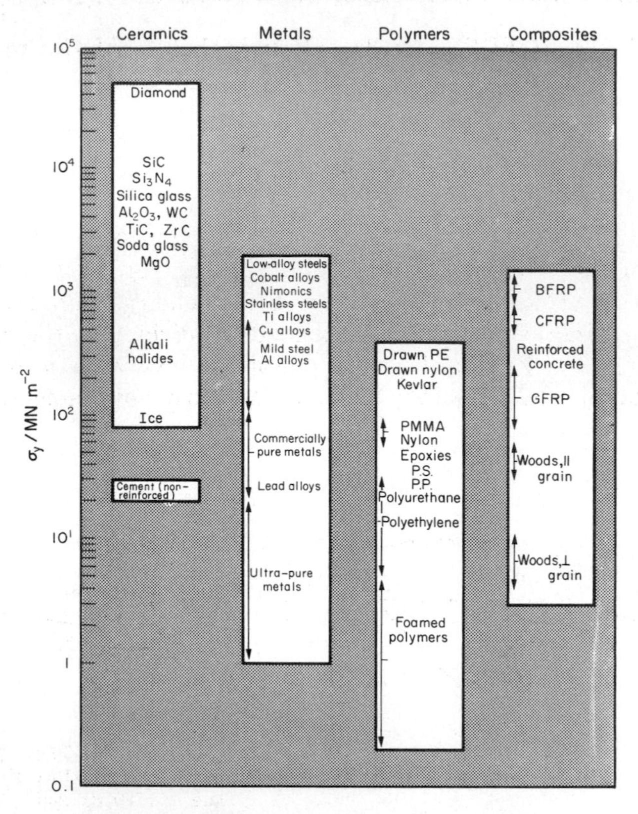

Fig. 8.12. Bar-chart of data for yield strength, σ_y.

10^6: from about $0.1\,\text{MN m}^{-2}$ (for polystyrene foams) to nearly $10^5\,\text{MN m}^{-2}$ (for diamond).

Most ceramics have enormous yield stresses. In a tensile test, at room temperature, ceramics almost all fracture long before they yield: this is because their fracture toughness, which we will discuss later, is very low. Because of this, you cannot measure the yield strength of a ceramic by using a tensile test. Instead, you have to use a test which somehow suppresses fracture: a compression test, for instance. The best and easiest is the hardness test: the data shown here are obtained from hardness tests, which we shall discuss in a moment.

Pure metals are very soft indeed, and have a high ductility. This is what, for centuries, has made them so attractive at first for jewellery and weapons, and then for other implements and structures: they can be *worked* to the shape that you want them in; furthermore, their ability to work-harden means that, after you have finished, the metal is much stronger than when you started. By alloying, the strength of metals can be further increased, though—in yield strength—the strongest metals still fall short of most ceramics.

Polymers, in general, have lower yield strengths than metals. The very strongest (and, at present, these are produced only in small quantities, and are expensive) barely reach the strength of aluminium alloys. They can be strengthened, however, by making composites out of them: GFRP has a strength only slightly inferior to aluminium, and CFRP is substantially stronger.

The hardness test

This consists of loading a pointed diamond or a hardened steel ball and pressing it into the surface of the material to be examined. The further into the material the "indenter" (as it is called) sinks, the *softer* is the material and the lower its yield strength. The *true hardness* is defined as the load (F) divided by the projected area of the "indent", A. (The Vickers hardness, H_v, unfortunately was, and still is, defined as F divided by the total surface area of the "indent". Tables are available to relate H to H_v.)

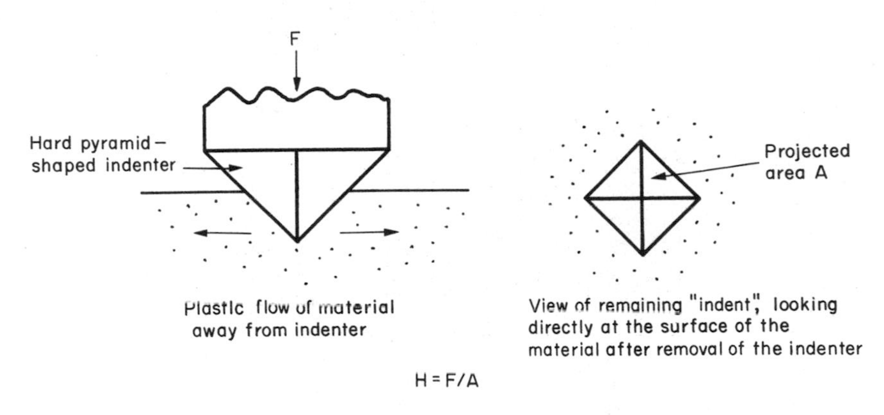

Fig. 8.13. The hardness test for yield strength.

The yield strength can be found from the relation (derived in Chapter 11)

$$H = 3\sigma_y \tag{8.8}$$

but a correction factor is needed for materials which work harden appreciably.

As well as being a good way of measuring the yield strengths of materials like ceramics, as we mentioned above, the hardness test is also a very simple and cheap *non-destructive test for* σ_y. There is no need to go to the expense of making tensile specimens, and the hardness indenter is so small that it scarcely damages the material. So it can be used for routine batch tests on materials to see if they are up to specification on σ_y without damaging them.

Further reading

K. J. Pascoe, *An Introduction to the Properties of Engineering Materials*, 3rd edition, Van Nostrand, 1978, Chap. 12.

C. J. Smithells, *Metals Reference Book*, 5th edition, Butterworths, 1976 (for data).

Revision of the terms mentioned in this chapter, and some useful relations

σ_n, *nominal stress*

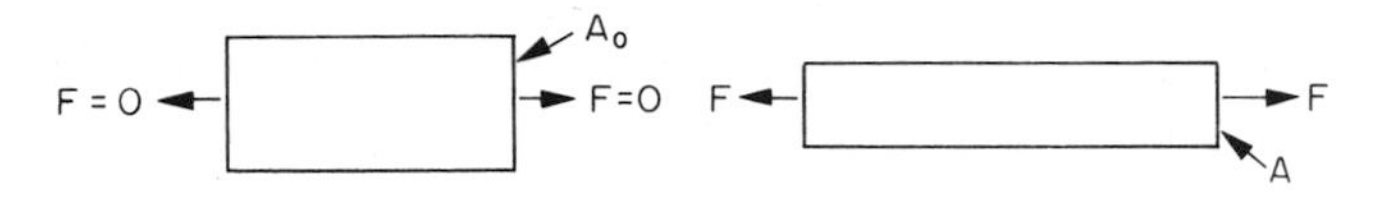

Fig. 8.14.

$$\sigma_n = F/A_0. \tag{8.9}$$

σ, *true stress*

$$\sigma = F/A. \tag{8.10}$$

ε_n, *nominal strain*

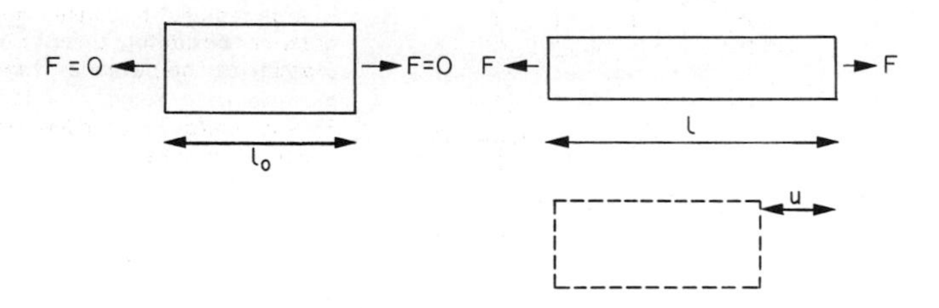

Fig. 8.15.

$$\varepsilon_n = \frac{u}{l_0}, \quad \text{or} \quad \frac{l - l_0}{l_0}, \quad \text{or} \quad \frac{l}{l_0} - 1. \tag{8.11}$$

Relations between σ_n, σ, and ε_n

Assuming constant volume (valid if $\nu = 0.5$ *or*, if not, plastic deformation $\gg$ elastic deformation):

$$A_0 l_0 = Al; \quad A_0 = \frac{Al}{l_0} = A(1 + \varepsilon_n). \tag{8.12}$$

Thus

$$\sigma = \frac{F}{A} = \frac{F}{A_0}(1 + \varepsilon_n) = \sigma_n(1 + \varepsilon_n). \tag{8.13}$$

ε, true strain and the relation between ε and ε_n

$$\varepsilon = \int_{l_0}^{l} \frac{dl}{l} = \ln\left(\frac{l}{l_0}\right). \tag{8.14}$$

Thus

$$\varepsilon = \ln(1+\varepsilon_n). \tag{8.15}$$

Small strain condition

For small ε_n

$$\varepsilon \approx \varepsilon_n, \quad \text{from} \quad \varepsilon = \ln(1+\varepsilon_n), \tag{8.16}$$

$$\sigma \approx \sigma_n, \quad \text{from} \quad \sigma = \sigma_n(1+\varepsilon_n). \tag{8.17}$$

Thus, when dealing with most *elastic* strains (but not in rubbers), it is immaterial whether ε or ε_n, or σ or σ_n, are chosen.

Energy

The energy expended in deforming a material *per unit volume* is given by the area under the stress–strain curve. For example,

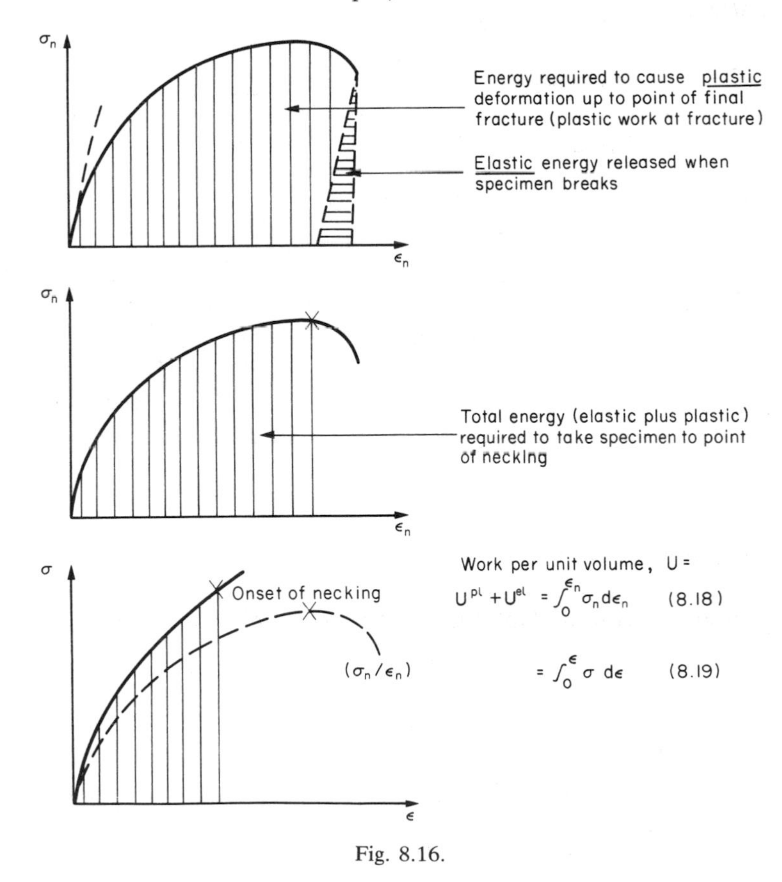

Fig. 8.16.

For *linear elastic strains*, and *only* linear elastic strains,

$$\frac{\sigma_n}{\varepsilon_n} = E, \quad \text{and} \quad U^{el} = \int \sigma_n \, d\varepsilon_n = \int \sigma_n \frac{d\sigma_n}{E} = \left\{\frac{\sigma_n^2}{2E}\right\}. \tag{8.20}$$

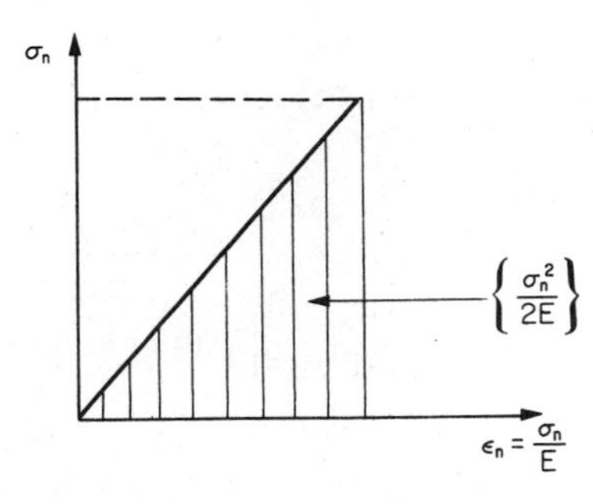

Fig. 8.17.

Elastic limit

In a tensile test, as the load increases, the specimen at first is strained *elastically*, that is reversibly. Above a limiting stress—the elastic limit—some of the strain is permanent; this is *plastic* deformation.

Yielding

The change from elastic to measurable plastic deformation.

Yield strength

The nominal stress at yielding. In many materials this is difficult to spot on the stress–strain curve and in such cases it is better to use a proof stress.

Proof stress

The stress which produces a permanent strain equal to a specified percentage of the specimen length. A common proof stress is one corresponding to 0.1% permanent strain.

Strain hardening (work-hardening)

The increase in stress needed to produce further strain in the plastic region. Each strain increment strengthens or hardens the material so that a larger stress is needed for further strain.

σ_{TS}, tensile strength (in old books, ultimate tensile strength, or UTS)

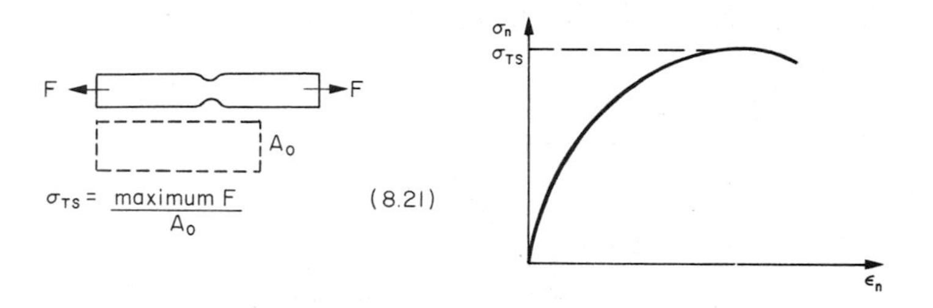

Fig. 8.18.

ε_f, strain after fracture, or tensile ductility

The permanent extension in length (measured by fitting the broken pieces together) expressed as a percentage of the original gauge length.

$$\left\{\frac{l_{\text{break}} - l_0}{l_0}\right\} \times 100. \tag{8.22}$$

Reduction in area at break

The maximum decrease in cross-sectional area at the fracture expressed as a percentage of the original cross-sectional area.

$$\left\{\frac{A_0 - A_{\text{break}}}{A_0}\right\} \times 100 \tag{8.23}$$

A

A_{break}

Fig. 8.19.

Strain after fracture and percentage reduction in area are used as measures of ductility, i.e. the ability of a material to undergo large plastic strain under stress before it fractures.

CHAPTER 9

DISLOCATIONS AND YIELDING IN CRYSTALS

Introduction

In the last chapter we examined data for the yield strengths exhibited by materials. But what would we expect? From our understanding of the structure of solids and the stiffness of the bonds between the atoms, can we estimate what the yield strength should be? A simple calculation (given in the next section) overestimates it grossly. This is because real crystals contain defects, *dislocations*, which move easily. When they move, the crystal deforms; the stress needed to move them is the yield strength. Dislocations are the *carriers* of deformation, much as electrons are the carriers of charge.

The strength of a perfect crystal

As we showed in Chapter 6 (on the modulus), the slope of the interatomic force–distance curve at the equilibrium separation is proportional to Young's modulus

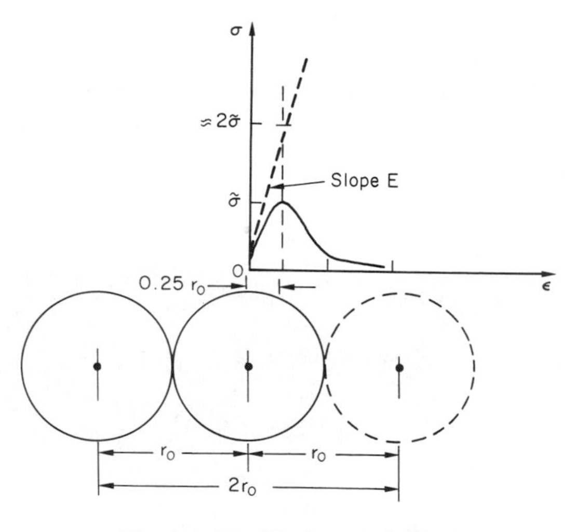

Fig. 9.1. The ideal strength, $\tilde{\sigma}$.

E. Interatomic forces typically drop off to negligible values at a distance of separation of the atom centres of $2r_0$. The maximum in the force–distance curve is typically reached at $1.25r_0$ separation, and if the stress applied to the material is sufficient to exceed this maximum force per *bond*, fracture is bound to occur. We will denote the stress at which this bond rupture takes place by $\tilde{\sigma}$, the *ideal strength*; a material cannot be stronger than this. From Fig. 9.1

$$\sigma = E\varepsilon,$$

$$2\tilde{\sigma} \approx E\frac{0.25r_0}{r_0} \approx \frac{E}{4},$$

$$\tilde{\sigma} \approx \frac{E}{8}. \tag{9.1}$$

More refined estimates of $\tilde{\sigma}$ are possible, using real interatomic potentials (Chapter 4): they give about $E/15$ instead of $E/8$.

Let us now see whether materials really show this strength. The bar-chart (Fig. 9.2) shows values of σ_y/E for materials. The heavy broken line at the top is drawn at the level $\sigma/E = 1/15$. Glasses, and some ceramics, lie close to this line—they exhibit their ideal strength, and we could not expect them to be stronger than this. Most polymers, too, lie near the line—although they have low yield strengths, these are low because the *moduli* are low.

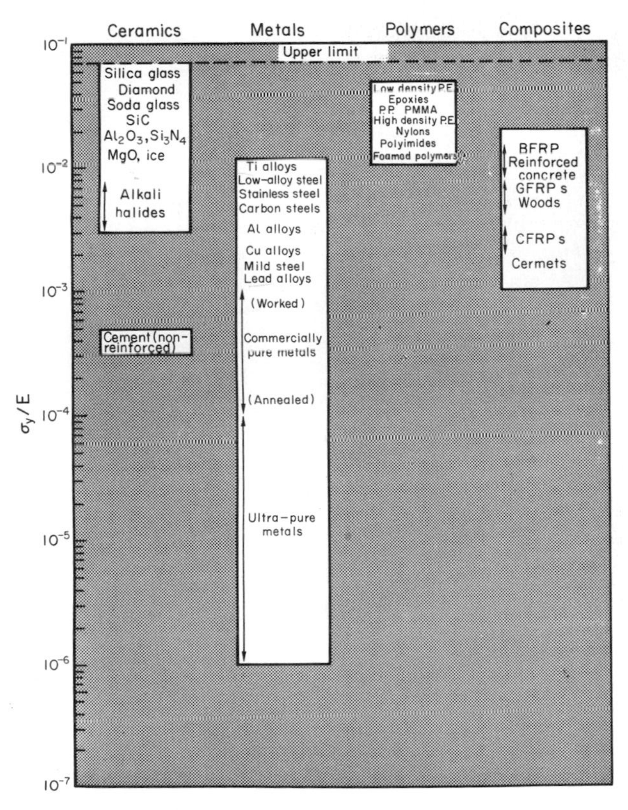

Fig. 9.2. Bar-chart of data for normalized yield strength, σ_y/E.

All metals, on the other hand, have yield strengths far below the levels predicted by our calculation—as much as a factor of 10^5 smaller. Even ceramics, many of them, yield at stresses which are as much as a factor of 10 below their ideal strength. Why is this?

Dislocations in crystals

In Chapter 5 we said that many important engineering materials (e.g. metals) were normally made up of crystals, and explained that a perfect crystal was an assembly of *atoms packed together in a regularly repeating pattern.*

But crystals (like everything in this world) are not perfect; they have *defects* in them. Just as the strength of a chain is determined by the strength of the weakest link, so the strength of a crystal—and thus of our material—is usually limited by the defects that are present in it. The *dislocation* is a particular type of defect that has the effect of allowing materials to deform plastically (that is, they yield) at stress levels that are much less than $\tilde{\sigma}$.

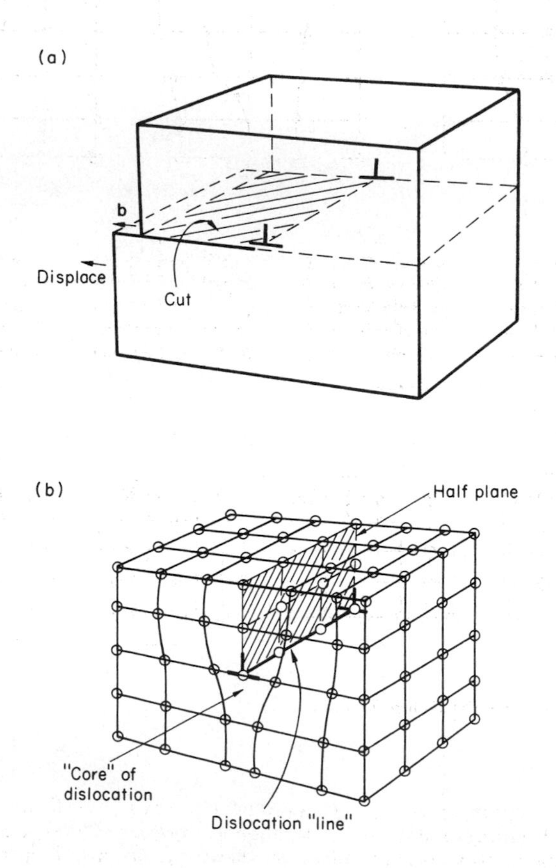

Fig. 9.3. An edge dislocation, (a) viewed from a continuum standpoint (i.e. ignoring the atoms) and (b) showing the positions of the atoms near the dislocation.

Figure 9.3(a) shows an *edge dislocation* from a continuum viewpoint (i.e. ignoring the atoms). Such a dislocation is made in a block of material by cutting the block up to the line marked ⊥—⊥, then displacing the material below the cut relative to that above by a distance b (the atom size) normal to the line ⊥—⊥, and finally gluing the cut-and-displaced surfaces back together. The result, on an atomic scale, is shown in the adjacent diagram (Fig. 9.3(b)); the material in the middle of the block now contains a *half-plane* of atoms, with its lower edge lying along the line ⊥—⊥: the *dislocation line.* This defect is called an edge dislocation because it is formed by the edge of the half-plane of atoms; and it is written briefly by using the symbol ⊥.

Dislocation motion produces plastic strain. Figure 9.4 shows how the atoms re-arrange as the dislocation moves through the crystal, and that, when one dislocation moves entirely through a crystal, the lower part is displaced under the upper by the distance b (called the Burgers vector). The same process is drawn, without the atoms, and using the symbol ⊥ for the position of the dislocation line, in Fig. 9.5. The way in which this dislocation works can be likened to the way in which a ballroom carpet can

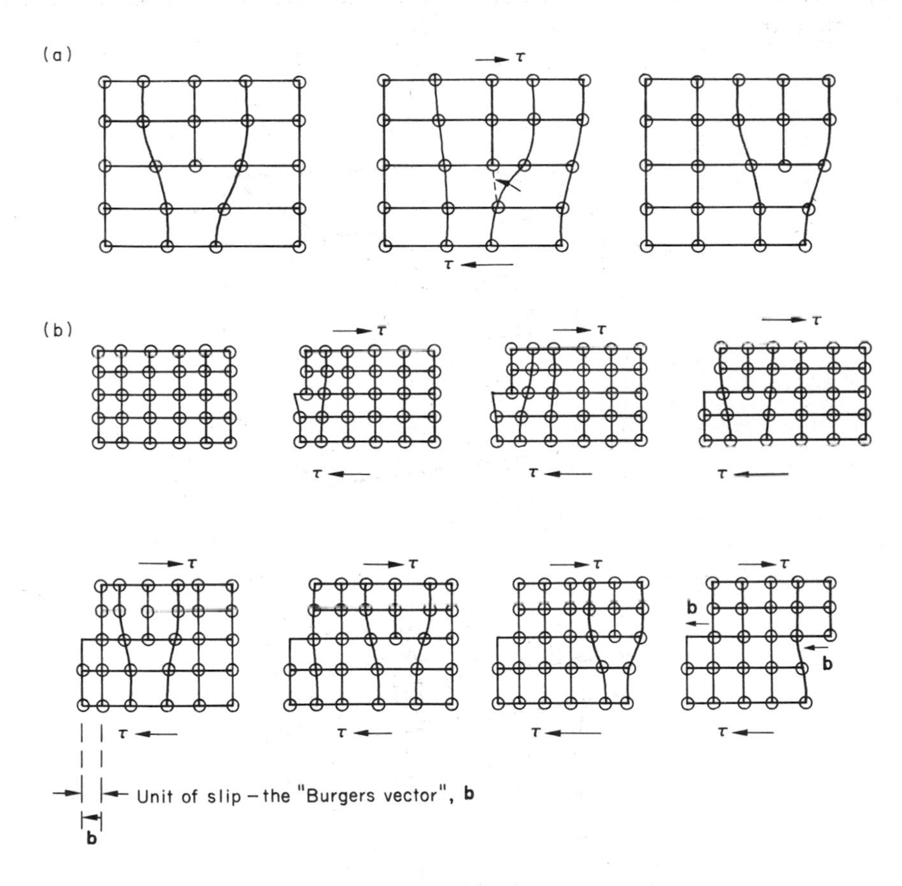

Fig. 9.4. How an edge dislocation moves through a crystal. (a) Shows how the atomic bonds at the centre of the dislocation break and reform to allow the dislocation to move. (b) Shows a complete sequence for the introduction of a dislocation into a crystal from the left-hand side, its migration through the crystal, and its expulsion on the right-hand side; this process causes the lower half of the crystal to slip by a distance b under the upper half.

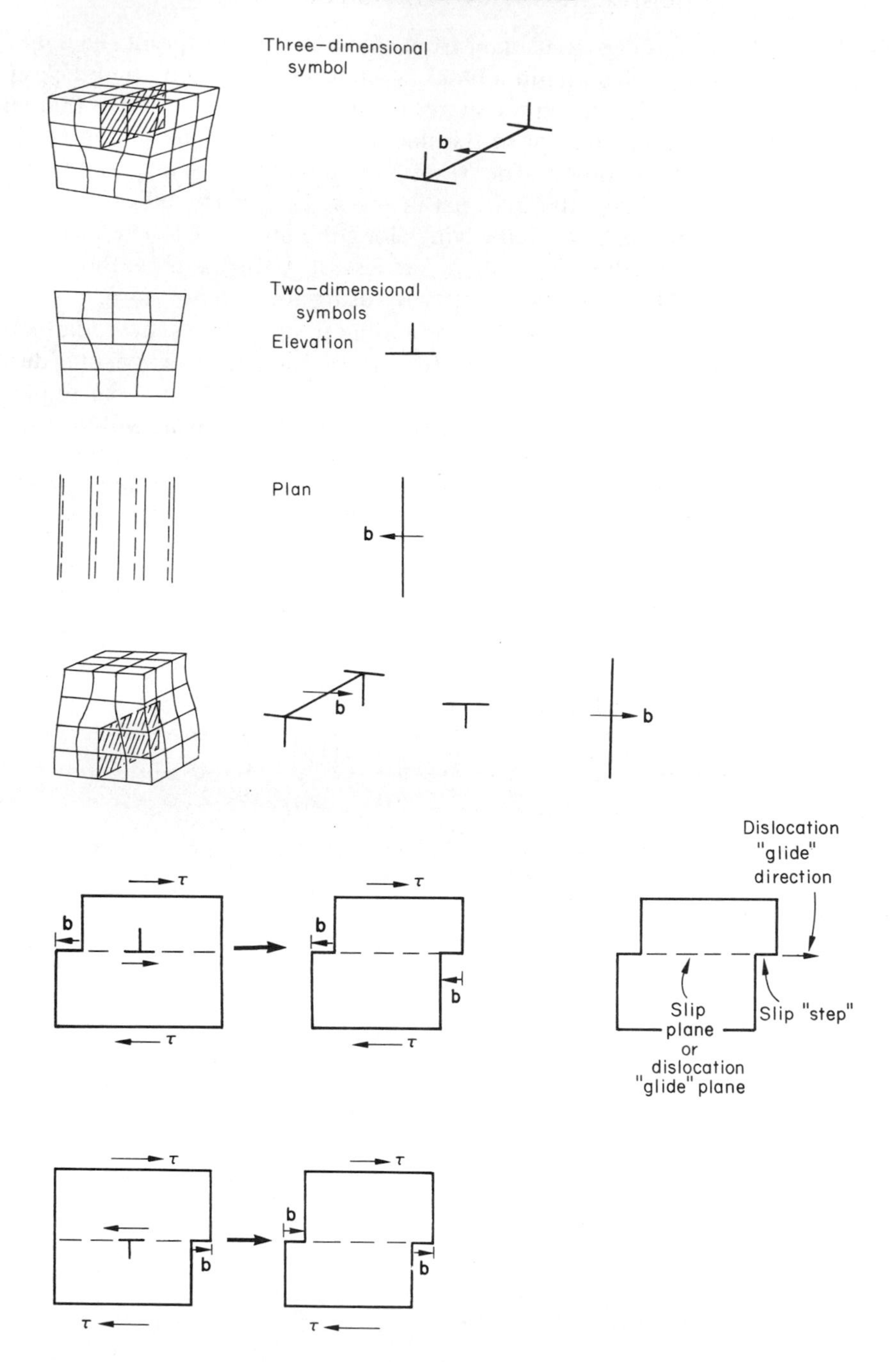

Fig. 9.5. Edge-dislocation conventions.

be moved across a large dance floor simply by moving rucks along the carpet—a very much easier process than pulling the whole carpet across the floor at one go.

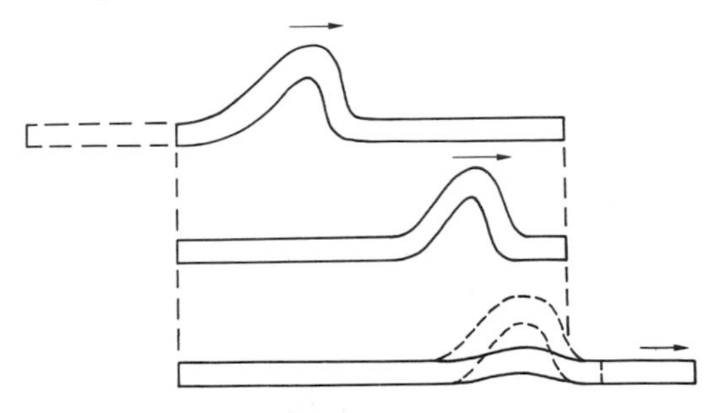

Fig. 9.6. The "carpet-ruck" analogy of an edge dislocation.

In making the edge dislocation of Fig. 9.3 we could, after making the cut, have displaced the lower part of the crystal under the upper part in a direction *parallel* to the bottom of the cut, instead of normal to it. Figure 9.7 shows the result; it, too, is a

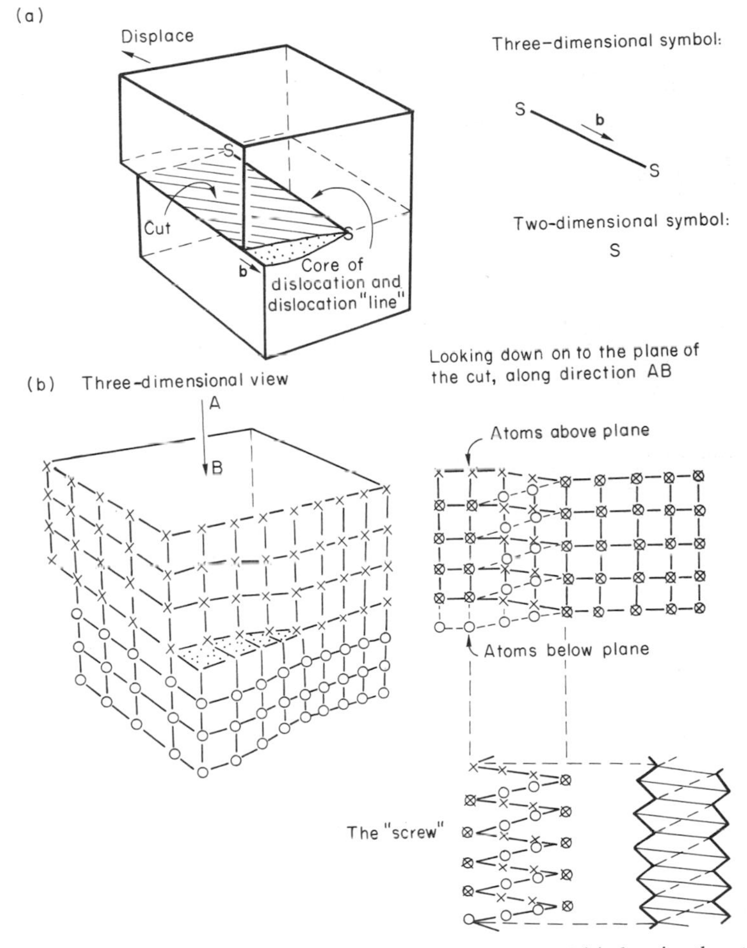

Fig. 9.7. A screw dislocation, (a) viewed from a continuum standpoint and (b) showing the atom positions.

dislocation, called a *screw dislocation* (because it converts the planes of atoms into a helical surface, or screw). Like an edge dislocation, it produces plastic strain when it moves (Figs. 9.8, 9.9, 9.10). Its geometry is a little more complicated but its properties are otherwise just like those of the edge. Any dislocation, in a real crystal, is either a screw or an edge; or can be thought of as little steps of each. Dislocations can be seen by electron microscopy. Figure 9.11 shows an example.

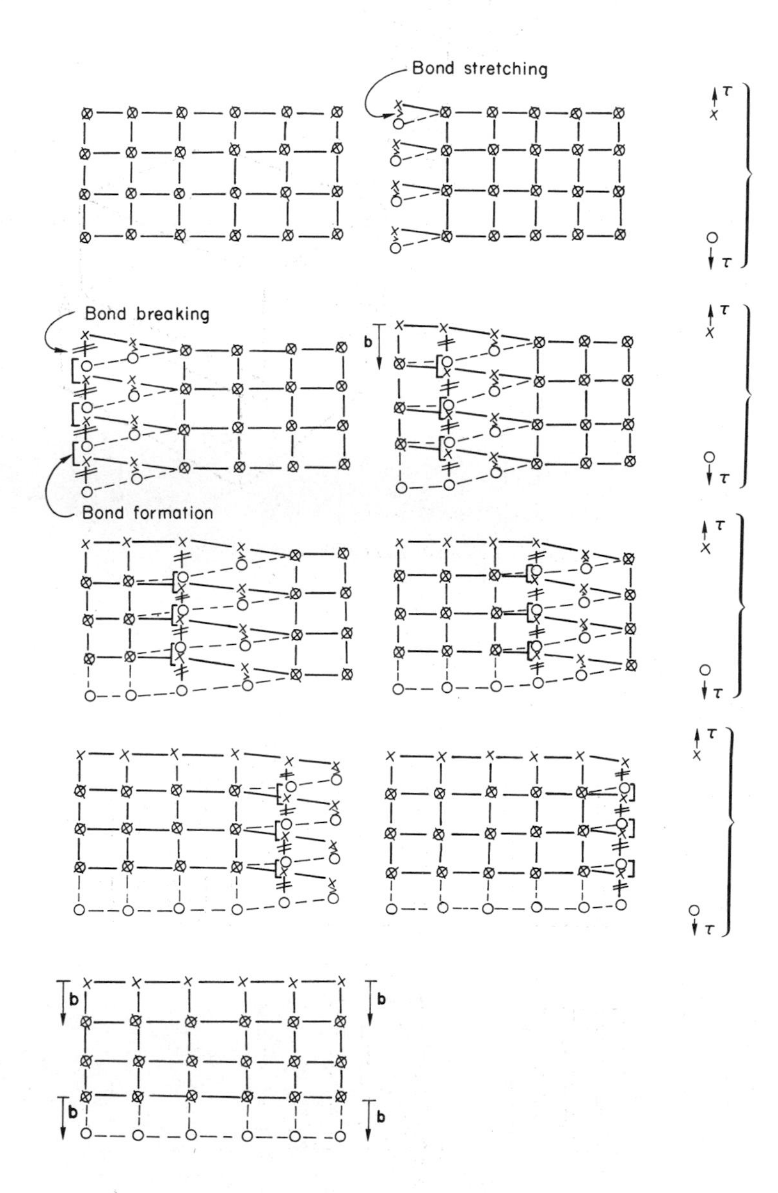

Fig. 9.8. Sequence showing how a screw dislocation moves through a crystal causing the lower half of the crystal (o) to slip by a distance b under the upper half (x).

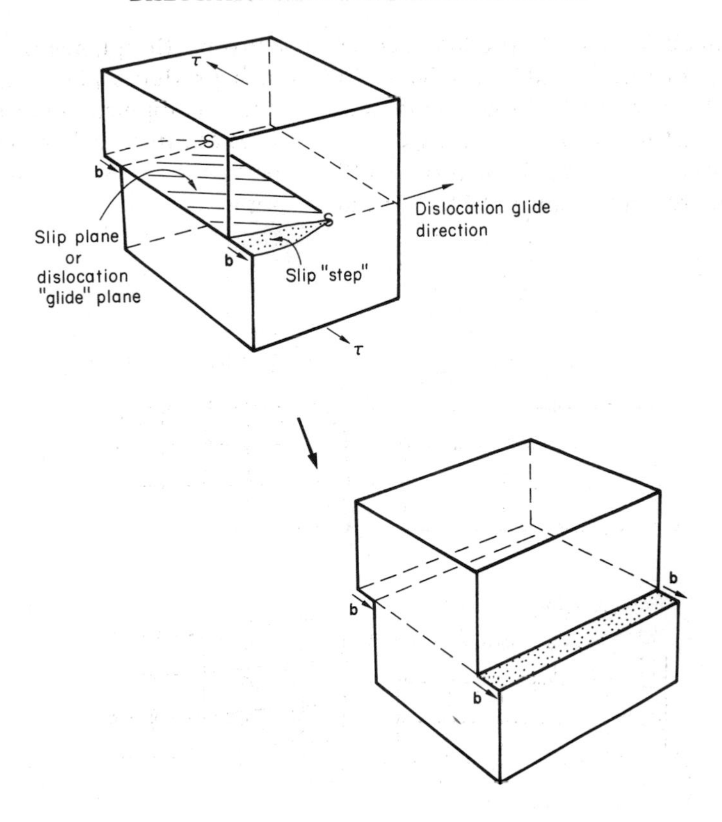

Fig. 9.9. Screw-dislocation conventions.

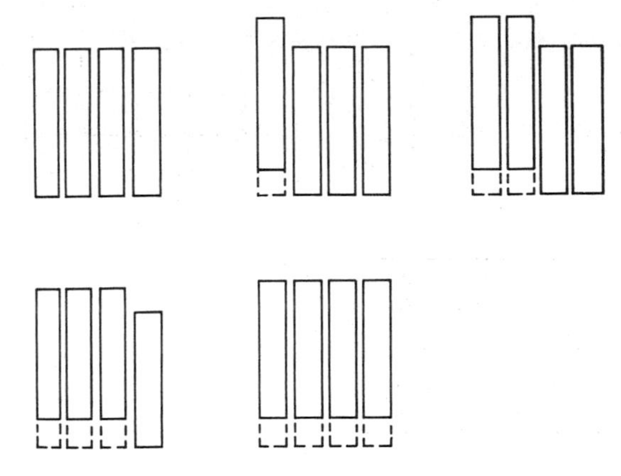

Fig. 9.10. The "planking" analogy of the screw dislocation. Imagine four planks resting side by side on a factory floor. It is much easier to slide them across the floor one at a time than all at the same time.

Fig. 9.11. An electron microscope picture of dislocation lines in stainless steel. The picture was taken by firing electrons through a very thin slice of steel about 100 nm thick. The dislocation lines here are only about 1000 atom diameters long because they have been "chopped off" where they meet the top and bottom surfaces of the thin slice. But a sugar-cube-sized piece of any engineering alloy contains about 10^5 km of dislocation line. (Courtesy of Dr. Peter Southwick.)

The force acting on a dislocation

A shear stress (τ) exerts a force on a dislocation, pushing it through the crystal. For yielding to take place, this force must be great enough to overcome the *resistance* to the motion of the dislocation. This resistance is due to intrinsic friction opposing dislocation motion, plus contributions from alloying or work-hardening; they are discussed in detail in the next chapter. Here we show that the magnitude of the force is τb per unit length of dislocation.

We prove this by a virtual work calculation. We equate the work done by the applied stress when the dislocation moves completely through the crystal to the work done against the force f opposing its motion (Fig. 9.12). The upper part is displaced relative to the lower by the distance b, and the applied stress does work $(\tau l_1 l_2) \times b$. In moving through the crystal, the dislocation travels a distance l_2, doing work against the resistance, f per unit length, as it does so; this work is $f l_1 l_2$. Equating the two gives

$$\tau b = f. \tag{9.2}$$

This result holds for any dislocation—edge, screw or a mixture of both.

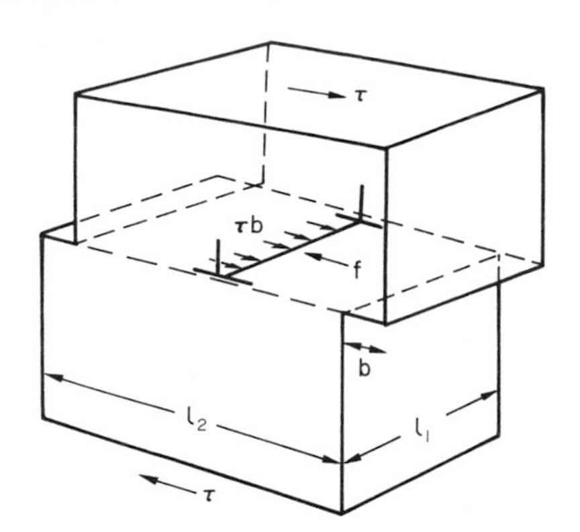

Fig. 9.12. The force acting on a dislocation.

Other properties of dislocations

There are two remaining properties of dislocations that are important in understanding the plastic deformation of materials. These are:

(a) Dislocations always glide on crystallographic planes, as we might imagine from our earlier drawings of edge-dislocation motion. In f.c.c. crystals, for example, the dislocations glide on $\{111\}$ planes, and therefore plastic shearing takes place on $\{111\}$ in f.c.c. crystals.

(b) The atoms near the core of a dislocation are displaced from their proper positions and thus have a higher energy. In order to keep the total energy as low as possible, the dislocation tries to be as short as possible—it behaves as if it had a *line tension*, T, like a rubber band. Very roughly, the strains at a dislocation core are of order 1/2; the stresses are therefore of order $G/2$ (Chapter 8) so the energy per unit volume of core is $G/8$. If we take the core radius to be equal to the atom size b, its volume, per unit length, is πb^2. The line tension is the energy per unit length (just as a surface tension is an energy per unit area), giving

$$T = \frac{\pi}{8} Gb^2 \approx \frac{Gb^2}{2}, \tag{9.3}$$

where G is the shear modulus. In absolute terms, T is small (we should need $\approx 10^8$ dislocations to hold an apple up) but it is large in relation to the size of a dislocation, and has an important bearing on the way in which obstacles obstruct the motion of dislocations.

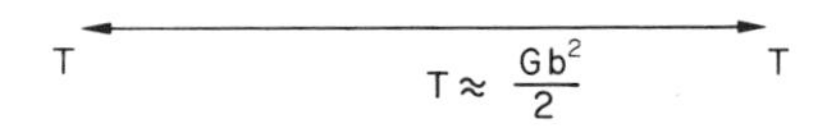

Fig. 9.13. The line tension in a dislocation.

We shall be looking in the next chapter at how we can use our knowledge of how dislocations work and how they behave in order to understand how materials deform plastically, and to help us design stronger materials.

Further reading

A. H. Cottrell, *The Mechanical Properties of Matter*, Wiley, 1964, Chap. 9.
D. Hull, *Introduction to Dislocations*, 2nd edition, Pergamon Press, 1975.
W. T. Read, Jr., *Dislocations in Crystals*, McGraw Hill, 1953.
J. P. Hirth and J. Lothe, *Theory of Dislocations*, McGraw Hill, 1968.

CHAPTER 10

STRENGTHENING METHODS, AND PLASTICITY OF POLYCRYSTALS

Introduction

We showed in the last chapter that:

(a) crystals contain dislocations;

(b) a shear stress τ, applied to the slip plane of a dislocation, exerts a force τb per unit length of the dislocation trying to push it forward;

(c) when dislocations move, the crystal deforms plastically—that is, it yields.

In this chapter we examine ways of increasing the resistance to motion of a dislocation; it is this which determines the *dislocation yield strength* of a single isolated crystal of a metal or a ceramic. But bulk engineering materials are aggregates of many crystals, or *grains*. To understand the plasticity of such an aggregate, we have to examine also how the individual crystals interact with each other. This lets us calculate the *polycrystal yield strength*—the quantity that enters engineering design.

Strengthening mechanisms

A crystal yields when the force τb (per unit length) exceeds f, the *resistance* (a force per unit length) opposing the motion of a dislocation. This defines the dislocation yield strength

$$\tau_y = \frac{f}{b}. \tag{10.1}$$

Most crystals have a certain *intrinsic* strength, caused by the bonds between the atoms which have to be broken and reformed as the dislocation moves. Covalent bonding, particularly, gives a very large *intrinsic lattice resistance*, f_i per unit length of dislocation. It is this that causes the enormous strength and hardness of diamond, and the carbides, oxides, nitrides and silicates which are used for abrasives and cutting tools. But pure metals are very soft: they have a very low lattice resistance. Then it is useful to increase f by *solid solution strengthening*, by *precipitate* or *dispersion* strengthening, or by *work-hardening*, or by any combination of the three. Remember, however, that there is an upper limit to the yield strength: it can never exceed the ideal strength (Chapter 9). In practice, only a few materials have strengths that even approach it.

Solid solution hardening

A good way of hardening a metal is simply to make it impure. Impurities go into solution in a solid metal just as sugar dissolves in tea. A good example is the addition of zinc to copper to make the *alloy* called brass. The zinc atoms replace copper atoms to form a *random substitutional solid solution.* At room temperature Cu will dissolve up to 30% Zn in this way. The Zn atoms are bigger than the Cu atoms, and, in squeezing into the Cu structure, generate stresses. These stresses "roughen" the slip plane, making it harder for dislocations to move; they increase the resistance f, and thereby increase the dislocation yield strength, τ_y (eqn. (10.1)). If the contribution to f given by the solid solution is f_{ss} then τ_y is increased by f_{ss}/b. In a solid solution of concentration C, the spacing of dissolved atoms on the slip plane (or on any other plane, for that matter) varies as $C^{-1/2}$; and the smaller the spacing, the "rougher" is the slip plane. As a result, τ_y increases about parabolically (i.e. as $C^{1/2}$) with solute concentration (Fig. 10.1). Single-phase brass, bronze, and stainless steels, and many other metallic alloys, derive their strength in this way.

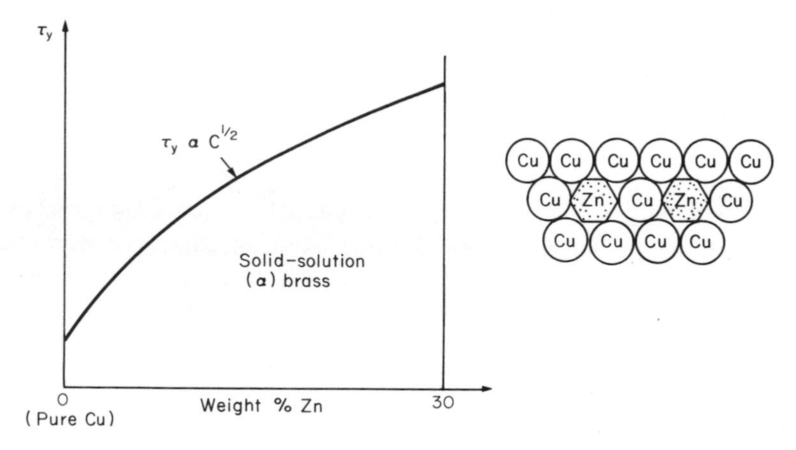

Fig. 10.1. Solid solution hardening.

Precipitate and dispersion strengthening

If an impurity (copper, say) is dissolved in a metal or ceramic (aluminium, for instance) at a high temperature, and the alloy is cooled to room temperature, the impurity may *precipitate* as small particles, much as sugar will crystallise from a saturated solution when it is cooled. An alloy of Al containing 4% Cu ("Duralumin"), treated in this way, gives very small, closely spaced precipitates of the hard compound $CuAl_2$. Most steels are strengthened by precipitates of carbides, obtained in this way.*

* The optimum precipitate is obtained by a more elaborate *heat treatment*: the alloy is *solution heat-treated* (heated to dissolve the impurity), *quenched* (cooled fast to room temperature, usually by dropping it into oil or water) and finally *tempered* or *aged* for a controlled time and at a controlled temperature (to cause the precipitate to form).

Small particles can be introduced into metals or ceramics in other ways. The most obvious is to mix a dispersoid (such as an oxide) into a powdered metal (aluminium and lead are both treated in this way), and then compact and sinter the mixed powders.

Either approach distributes small, hard particles in the path of a moving dislocation. Figure 10.2 shows how they obstruct its motion. The stress τ has to push the dislocation

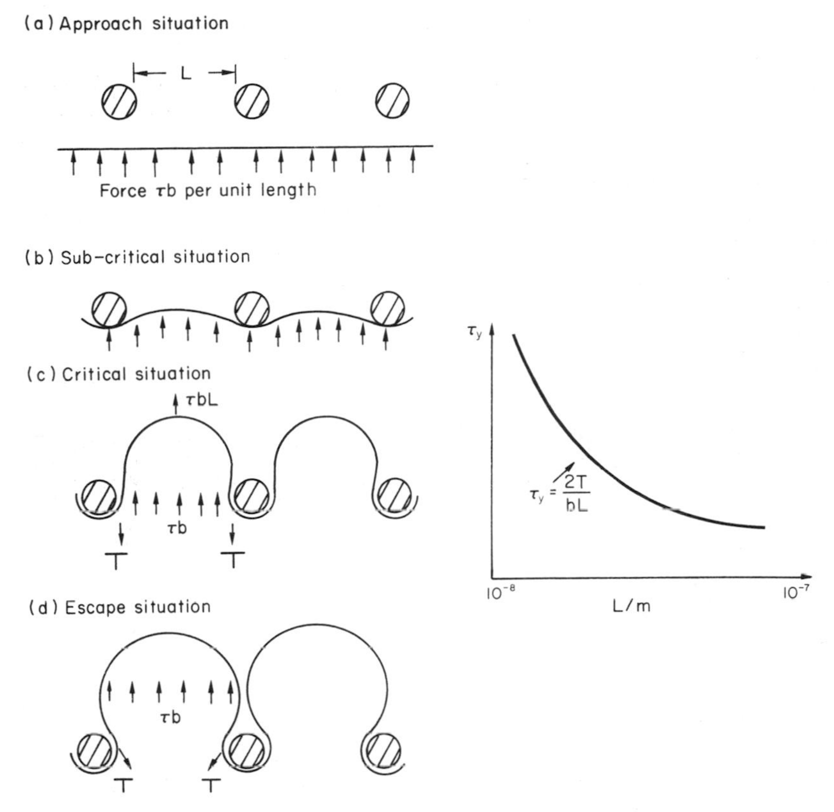

Fig. 10.2. How dispersed precipitates help prevent the movement of dislocations, and help prevent plastic flow of materials.

between the obstacles. It is like blowing up a balloon in a bird cage: a very large pressure is needed to bulge the balloon between the bars, though once a large enough bulge is formed, it can easily expand further. The *critical configuration* is the semicircular one (Fig. 10.2(c)): here the force τbL on one segment is just balanced by the force $2T$ due to the line tension, acting on either side of the bulge. The dislocation escapes (and yielding occurs) when

$$\tau_y = \frac{2T}{bL}. \tag{10.2}$$

The obstacles thus exert a resistance of $f_0 = 2T/L$. Obviously, the greatest hardening is produced by *strong, closely spaced* precipitates or dispersions (Fig. 10.2).

Work-hardening

When crystals yield, dislocations move through them. Most crystals have several slip planes: the f.c.c. structure, which slips on {111} planes (Chapter 5), has four, for example. Dislocations on these intersecting planes interact, and obstruct each other, and accumulate in the material.

The result is *work-hardening*: the steeply rising stress–strain curve after yield, shown in Chapter 8. All metals and ceramics work-harden. It can be a nuisance: if you want to roll thin sheet, work-hardening quickly raises the yield strength so much that you have to stop and *anneal* the metal (heat it up to remove the accumulated dislocations) before you can go on. But it is also useful: it is a potent strengthening method, which can be added to the other methods to produce strong materials.

The analysis of work-hardening is difficult. Its contribution $f_{\omega h}$ to the total dislocation resistance f is considerable and increases with strain (Fig. 10.3).

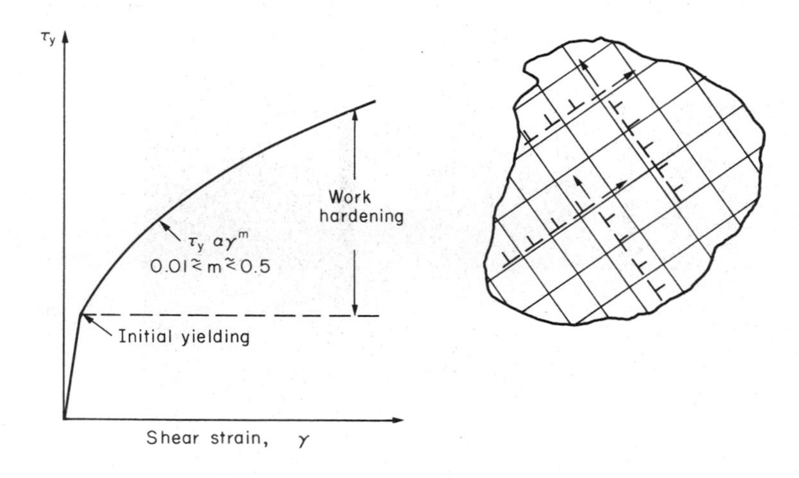

Fig. 10.3. Collision of dislocations leads to work-hardening.

The dislocation yield strength

It is adequate to assume that the strengthening methods contribute in an additive way to the strength. Then

$$\tau_y = \frac{f_i}{b} + \frac{f_{ss}}{b} + \frac{f_0}{b} + \frac{f_{\omega h}}{b}. \tag{10.3}$$

Strong materials either have a high intrinsic strength, f_i (like diamond), or they rely on the superposition of solid solution strengthening f_{ss}, obstacles f_0 and work-hardening $f_{\omega h}$ (like high-tensile steels). But before we can use this information, one problem remains: we have calculated the yield strength of an *isolated crystal* in *shear*. We want the yield strength of a *polycrystalline aggregate* in *tension*.

Yield in polycrystals

The crystals, or *grains*, in a polycrystal fit together exactly but their crystal orientations differ (Fig. 10.4). Where they meet, at *grain boundaries*, the crystal structure is disturbed, but the atomic bonds across the boundary are numerous and strong enough that the boundaries do not usually weaken the material.

Let us now look at what happens when a polycrystalline component begins to yield (Fig. 10.5). Slip begins in grains where there are slip planes as nearly parallel to τ as possible, e.g. grain (1). Slip later spreads to grains like (2) which are not so favourably oriented, and lastly to the worst oriented grains like (3). Yielding does not take place all at once, therefore, and there is no sharp polycrystalline yield point on the stress–strain curve. Further, gross (total) yielding does not occur at the dislocation-yield

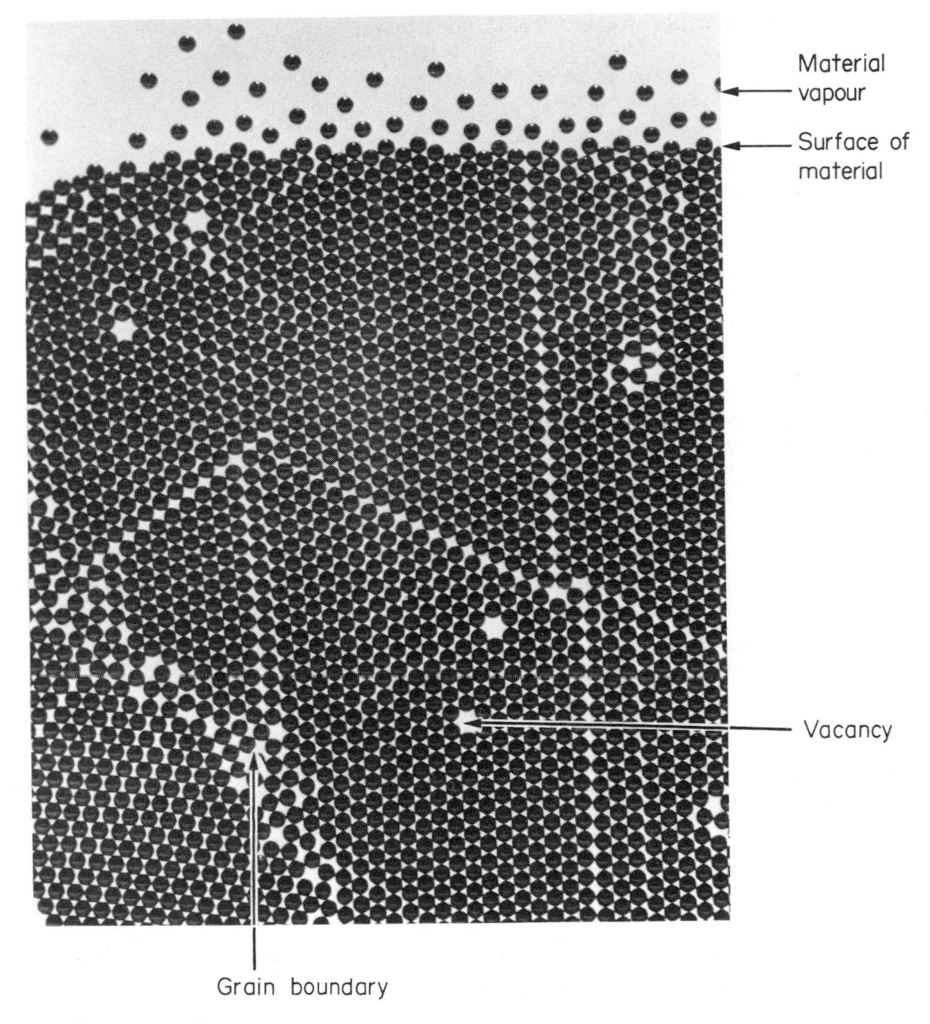

Fig. 10.4. Ball bearings can be used to simulate how atoms are packed together in solids. Our photograph shows a ball-bearing model set up to show what the *grain boundaries* look like in a polycrystalline material. The model also shows up another type of defect—the vacancy—which is caused by a missing atom.

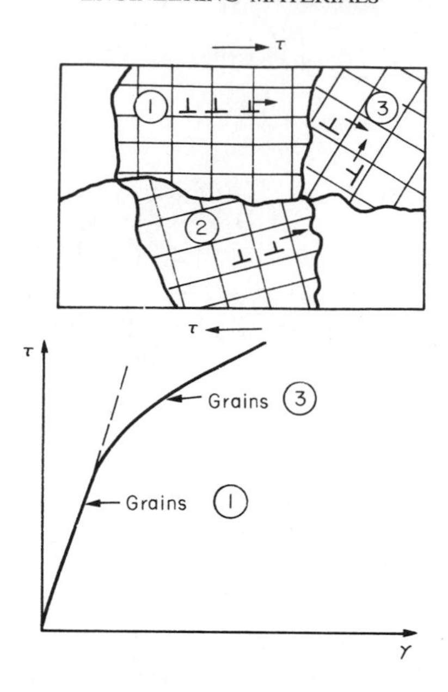

Fig. 10.5. The progressive nature of yielding in a polycrystalline material.

strength τ_y, because not all the grains are oriented favourably for yielding. The gross-yield strength is higher, by a factor called the Taylor factor, which is calculated (with difficulty) by averaging the stress over all possible slip planes; it is close to 1.5.

But we want the tensile yield strength, σ_y. A tensile stress σ creates a shear stress in the material that has a maximum value of $\tau = \sigma/2$. (We show this in Chapter 11 where we resolve the tensile stress onto planes within the material.) To calculate σ_y from τ_y, we combine the Taylor factor with the resolution factor to give

$$\sigma_y = 3\tau_y. \tag{10.4}$$

σ_y is the quantity we want: the yield strength of bulk, polycrystalline solids. It is larger than the dislocation shear strength τ_y (by the factor 3) but is proportional to it. So all the statements we have made about increasing τ_y apply unchanged to σ_y.

A whole science of alloy design for high strength has grown up in which alloys are blended and heat-treated to achieve maximum τ_y. Important components that are strengthened in this way range from lathe tools ("high-speed" steels) to turbine blades ("Nimonic" alloys based on nickel). We shall have more to say about strong solids when we come to look at how materials are *selected* for a particular job. But first we must return to a discussion of plasticity at the non-atomistic, or continuum, level.

Further reading

A. H. Cottrell, *The Mechanical Properties of Matter*, Wiley, 1964, Chap. 9.
R. W. K. Honeycombe, *The Plastic Deformation of Metals*, Arnold, 1968.

CHAPTER 11

CONTINUUM ASPECTS OF PLASTIC FLOW

Introduction

Plastic flow occurs by shear. Dislocations move when the shear stress on the slip plane exceeds *the dislocation yield strength* τ_y of a single crystal. If this is averaged over all grain-orientations and slip planes, it can be related to *the tensile yield strength* σ_y of a polycrystal by $\sigma_y = 3\tau_y$ (Chapter 10). But in solving problems of plasticity, it is more useful to define *the shear yield strength* k of a polycrystal. It is equal to $\sigma_y/2$, and differs from τ_y because it is an average shear-resistance over all orientations of slip plane. When a structure is loaded, the planes on which shear will occur can often be identified or guessed, and the collapse load calculated approximately by requiring that the stress exceed k on these planes.

In this chapter we show that $k = \sigma_y/2$, and use k to relate the hardness to the yield strength of a solid. We then examine tensile instabilities which appear in the drawing of metals and polymers.

The onset of yielding and the shear yield strength, k

A tensile stress applied to a piece of material will create a shear stress at an angle to the tensile stress. Let us examine the stresses in more detail. Resolving forces in Fig. 11.1 gives the shearing force as

$$F \sin \theta.$$

The area over which this force acts in shear is

$$\frac{A}{\cos \theta}$$

and thus the shear stress, τ, is

$$\tau = \frac{F \sin \theta}{A/\cos \theta} = \frac{F}{A} \sin \theta \cos \theta$$

$$= \sigma \sin \theta \cos \theta. \tag{11.1}$$

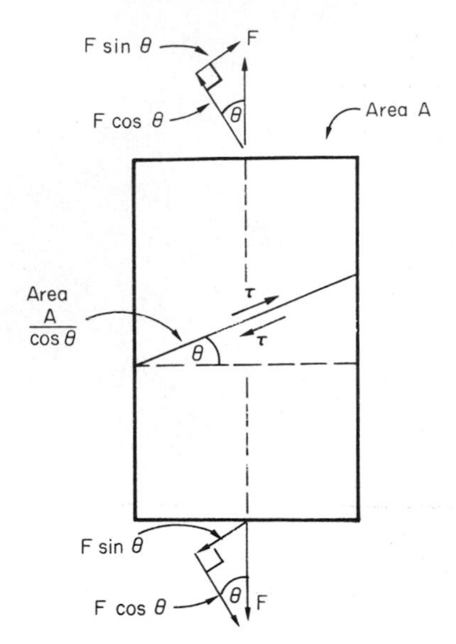

Fig. 11.1. A tensile stress, F/A, produces a shear stress, τ, on an inclined plane in the stressed material.

If we plot this against θ as in Fig. 11.2 we find a maximum τ at $\theta = 45°$ to the tensile axis. This means that the *highest value of the shear stress is found at* 45° *to the tensile axis, and has a value of* $\sigma/2$.

Now, from what we have said in Chapters 9 and 10, if we are dealing with a single crystal, the crystal will *not* in fact slip on the 45° plane—it will slip on the nearest lattice plane to the 45° plane on which dislocations can glide (Fig. 11.3). In a polycrystal, neighbouring grains each yield on their nearest-to-45° slip planes. On a microscopic scale, slip occurs on a zig-zag path; but the *average* slip path is at 45° to the tensile

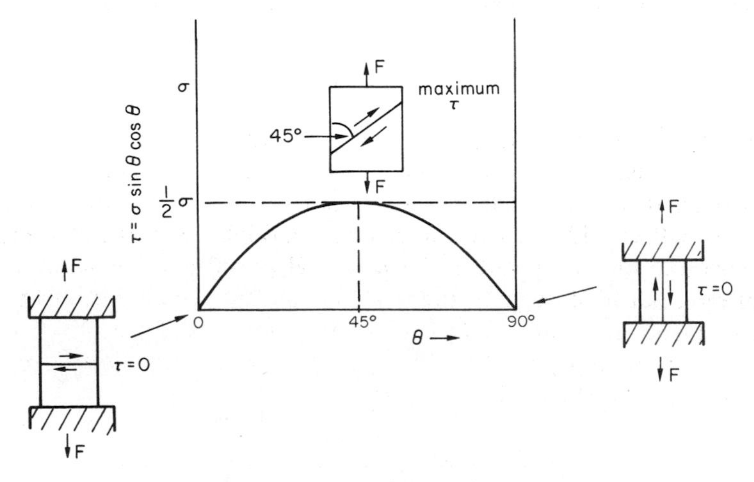

Fig. 11.2. Shear stresses in a material have their maximum value on planes at 45° to the tensile axis.

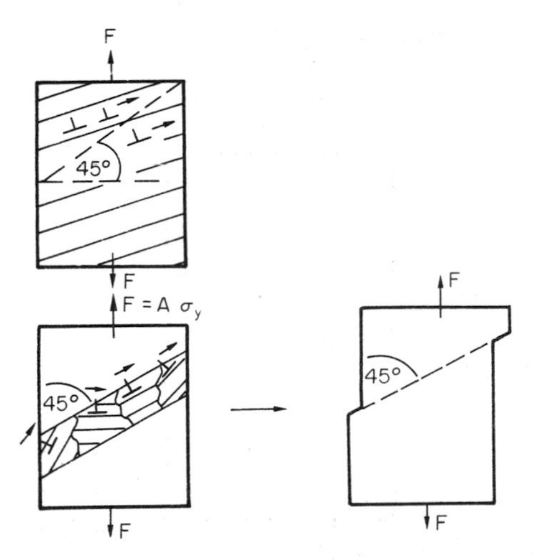

Fig. 11.3. In a polycrystalline material the *average* slip path is at 45° to the tensile axis.

axis. The shear stress on this plane when yielding occurs is therefore $\tau = \sigma_y/2$, and we define this as the shear yield strength k:

$$k = \sigma_{y/2}. \tag{11.2}$$

Example: *Approximate calculation of the hardness of solids.* This concept of shear yielding—where we ignore the details of the grains in our polycrystal and treat the material as a *continuum*—is useful in many respects. For example, we can use it to calculate the loads that would make our material yield for all sorts of quite complicated geometries.

A good example is the problem of the *hardness indenter* that we referred to in the hardness test in Chapter 8. Then, we stated that the hardness

$$H = \frac{F}{A} = 3\sigma_y$$

(with a correction factor for materials that work-harden appreciably—most do). For simplicity, let us assume that our material does not work-harden; so that as the indenter is pushed into the material, the yield strength does not change. Again, for simplicity, we will consider a two-dimensional model. (A real indenter, of course, is three-dimensional, but the result is, for practical purposes, the same.)

As we press a flat indenter into the material, shear takes place on the 45° planes of maximum shear stress shown in Fig. 11.4, at a value of shear stress equal to k. By equating the work done by the force F as the indenter sinks a distance u to the work done against k on the shear planes, we get:

$$Fu = 2 \times \frac{Ak}{\sqrt{2}} \times u\sqrt{2} + 2 \times Ak \times u + 4 \times \frac{Ak}{\sqrt{2}} \times \frac{u}{\sqrt{2}}.$$

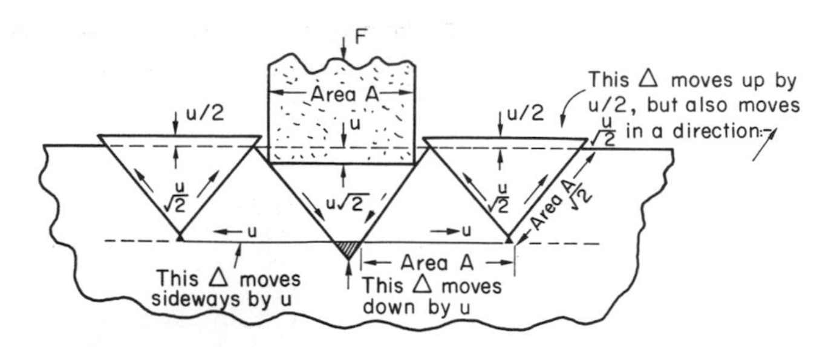

Fig. 11.4. The plastic flow of material under a hardness indenter—a simplified two-dimensional visualisation.

This simplifies to

$$F = 6Ak$$

from which

$$\frac{F}{A} = 6k = 3\sigma_y.$$

But F/A is the hardness, H; so

$$H = 3\sigma_y. \tag{11.3}$$

(Strictly, shear occurs not just on the shear planes we have drawn, but on a myriad of 45° planes near the indenter. If our assumed geometry for slip is wrong it can be shown rigorously by a theorem called the *upper-bound* theorem that the value we get for F at yield—the so-called "limit" load—is always on the high side.)

Similar treatments can be used for all sorts of two-dimensional problems: for calculating the plastic collapse load of structures of complex shape, and for analysing metal-working processes like forging, rolling and sheet drawing.

Plastic instability: necking in tensile loading

We now turn to the other end of the stress–strain curve and explain why, in tensile straining, materials eventually start to *neck*, a name for *plastic instability*. It means that flow becomes localised across one section of the specimen or component, as shown in Fig. 11.5, and (if straining continues) the material fractures there. Plasticine necks readily; chewing gum is very resistant to necking.

We analyse the instability by noting that if a force F is applied to the end of the specimen of Fig. 11.5, then any section must carry this load. But is it capable of doing so? Suppose one section deforms a little more than the rest, as the figure shows. Its section is less, and the stress in it is therefore larger than elsewhere. If work-hardening has raised the yield strength enough, the reduced section can still carry the force F; but if it has not, plastic flow will become localised at the neck and the specimen will fail there. Any section of the specimen can carry a force $A\sigma$, where A is its area, and σ its current strength. If $A\sigma$ increases with strain, the specimen is stable. If it decreases, it is

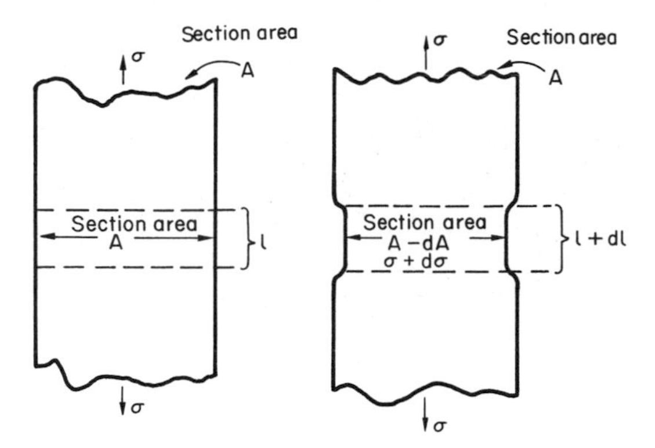

Fig. 11.5. The formation of a neck in a bar of material which is being deformed plastically.

unstable and will neck. The critical condition for the start of necking is that

$$A\sigma = F = \text{constant}.$$

Then

$$A\,\mathrm{d}\sigma + \sigma\,\mathrm{d}A = 0$$

or

$$\frac{\mathrm{d}\sigma}{\sigma} = -\frac{\mathrm{d}A}{A}.$$

But volume is conserved during plastic flow, so

$$-\frac{\mathrm{d}A}{A} = \frac{\mathrm{d}l}{l} = \mathrm{d}\varepsilon$$

(prove this by differentiating $Al = \text{constant}$). So

$$\frac{\mathrm{d}\sigma}{\sigma} = \mathrm{d}\varepsilon$$

or

$$\frac{\mathrm{d}\sigma}{\mathrm{d}\varepsilon} = \sigma. \tag{11.4}$$

This equation is given in terms of true stress and true strain. As we said in Chapter 8, tensile data are usually given in terms of nominal stress and strain. From Chapter 8:

$$\sigma = \sigma_n(1+\varepsilon_n),$$
$$\varepsilon = \ln(1+\varepsilon_n).$$

If these are differentiated and substituted into the necking equation we get

$$\frac{\mathrm{d}\sigma_n}{\mathrm{d}\varepsilon_n} = 0. \tag{11.5}$$

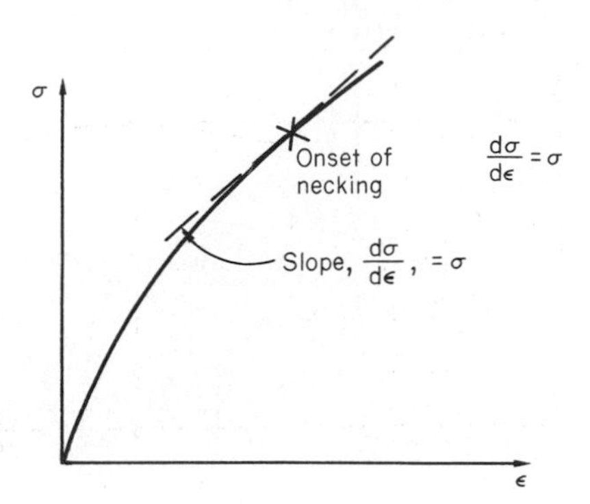

Fig. 11.6. The condition for necking.

In other words, on the point of instability, the *nominal* stress–strain curve is at its maximum as we know experimentally from Chapter 8.

To see what is going on physically, it is easier to return to our first condition. At low stress, if we make a little neck, the material in the neck will work-harden and will be able to carry the extra stress it has to stand because of its smaller area; load will therefore be continuous, and the material will be stable. At high stress, the *rate of work-hardening* is less as the true stress–true strain curve shows: i.e. the slope of the σ/ε curve is less. Eventually, we reach a point at which, when we make a neck, the work-hardening is only *just* enough to stand the extra stress. This is the point of necking, with

$$\frac{d\sigma}{d\varepsilon} = \sigma.$$

At still higher *true* stress, $d\sigma/d\varepsilon$, the rate of work-hardening decreases further, becoming insufficient to maintain stability—the extra stress in the neck can no longer be accommodated by the work-hardening produced by making the neck, and the neck grows faster and faster, until final fracture takes place.

Consequences of plastic instability

Plastic instability is very important in processes like deep drawing sheet metal to form car bodies, cans, etc. Obviously we must ensure that the materials and press designs are chosen carefully to *avoid* instability.

Mild steel is a good material for deep drawing in the sense that it flows a great deal before necking starts. It can therefore be drawn very deeply without breaking (Fig. 11.7).

Aluminium alloy is much less good (Fig. 11.8)—it can only be drawn a little before instabilities form. Pure aluminium is not nearly as bad, but is much too *soft* to use for most applications.

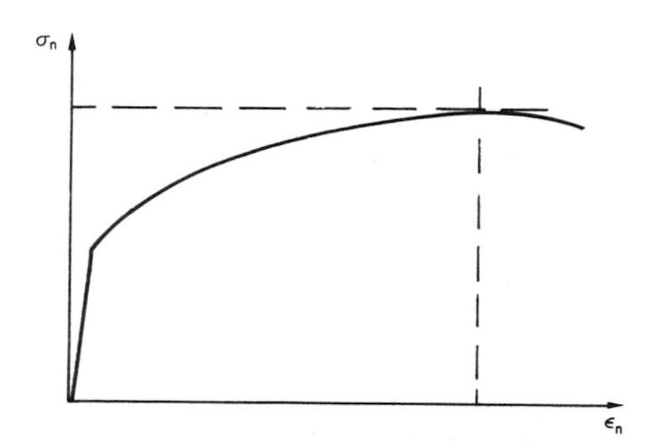

Fig. 11.7. Mild steel can be drawn out a lot before it fails by necking.

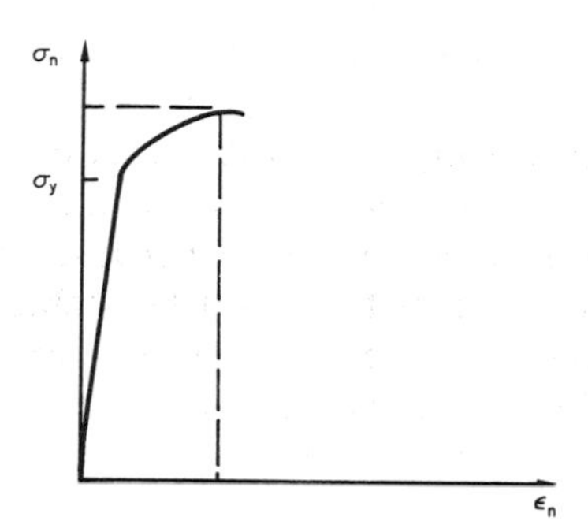

Fig. 11.8. Aluminium alloy quickly necks when it is drawn out.

Polythene shows a kind of necking that does *not* lead to fracture. Figure 11.9 shows its σ_n/ε_n curve. At quite low stress

$$\frac{d\sigma_n}{d\varepsilon_n}$$

becomes zero and necking begins. However, the neck never becomes *unstable*—it simply grows in length—because at high strain the material work-hardens considerably, and is able to support the increased stress at the reduced cross-section of the neck. This odd behaviour is caused by the lining up of the polymer chains in the neck along the direction of the neck—and for this sort of reason *drawn* (i.e. "fully necked") polymers can be made to be very strong indeed—much stronger than the undrawn polymers.

Finally, mild steel can sometimes show an instability like that of polythene. If the steel is annealed, the stress/strain curve looks like that in Fig. 11.10. A stable neck, called a Lüders Band, forms and propagates (as it did in polythene) without causing fracture because the strong work-hardening of the later part of the stress/strain curve

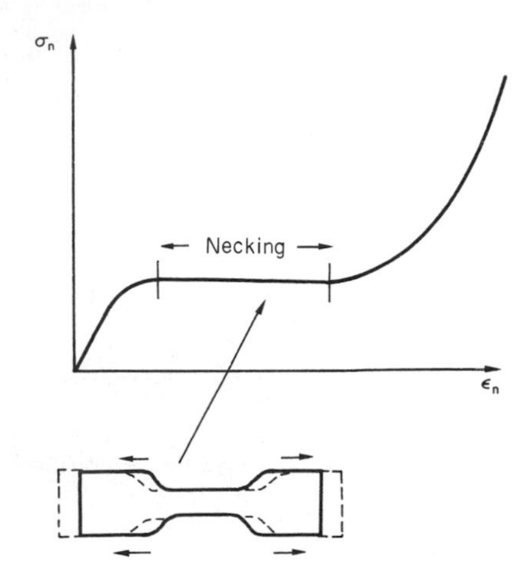

Fig. 11.9. Polythene forms a *stable* neck when it is drawn out; drawn polythene is very strong.

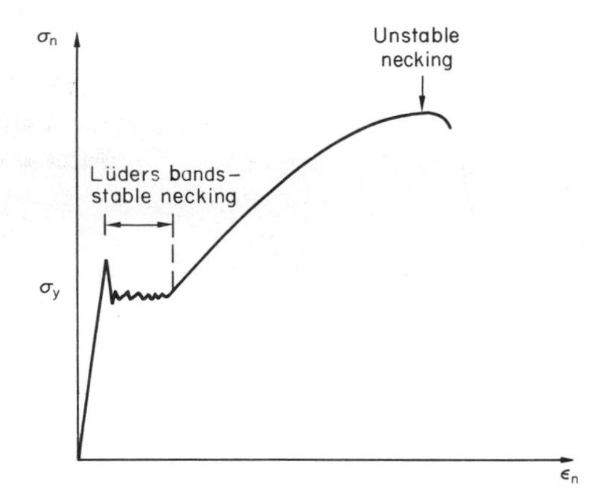

Fig. 11.10. Mild steel often shows both stable and unstable necks.

prevents this. Lüders Bands are a problem when sheet steel is pressed because they give lower precision and disfigure the pressing.

Further reading

A. H. Cottrell, *The Mechanical Properties of Matter*, Wiley, 1964, Chap. 10.
C. R. Calladine, *Engineering Plasticity*, Pergamon Press, 1969.
W. A. Backofen, *Deformation Processing*, Addison-Wesley, 1972.
R. Hill, *The Mathematical Theory of Plasticity*, Oxford University Press, 1950.

CHAPTER 12

CASE STUDIES IN YIELD-LIMITED DESIGN

Introduction

We now examine three applications of our understanding of plasticity. The first (material selection for a spring) requires that there be *no plasticity whatever.* The second (material selection for a pressure vessel) typifies plastic design of a large structure. It is unrealistic to expect no plasticity: there will always be some, at bolt holes, loading points, or changes of section. The important thing is that yielding should not spread entirely through any section of the structure—that is, that *plasticity must not become general.* Finally, we examine an instance (the rolling of metal strip) in which yielding is deliberately induced, to give *large-strain plasticity.*

CASE STUDY 1: ELASTIC DESIGN: MATERIALS FOR SPRINGS

Springs come in many shapes and sizes. Almost always, they are made of metal, the usual materials being as shown in Table 12.1.

TABLE 12.1
MATERIALS FOR SPRINGS

	E/GN m^{-2}	σ_y/MN m^{-2}	σ_y/E
Brass (cold-rolled)	120	638	5.32×10^{-3}
Bronze (cold-rolled)		640	5.33×10^{-3}
Phosphor bronze		770	6.43×10^{-3}
Beryllium copper		1380	11.5×10^{-3}
Spring steel	200	1300	6.5×10^{-3}
Stainless steel (cold-rolled)		1000	5.0×10^{-3}
Nimonic (high-temp. spring)		614	3.08×10^{-3}

At first sight, the modulus of the spring material might seem to be its most important single property; but, as you can see from the table, there is not much range of moduli among these seven common spring materials, and none have particularly high moduli. What, then, makes these materials good for springs? And, having understood that, could we suggest any new materials which might be used?

The leaf spring

We shall consider the particular case of the leaf spring. Even leaf springs can take many different forms, but all of them are basically small elastic beams loaded in bending. A rectangular section elastic beam, simply supported at both ends, loaded centrally with a force F, deflects by an amount

$$\delta = \frac{Fl^3}{4Ebt^3} \tag{12.1}$$

ignoring self-weight (Fig. 12.1).

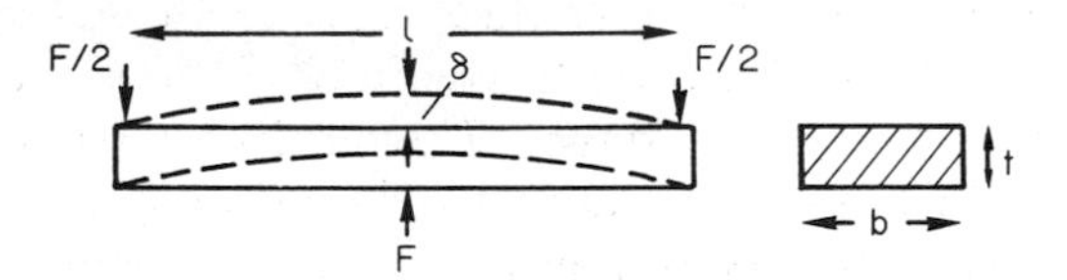

Fig. 12.1. A leaf spring under load.

Figure 12.2 shows that the *stress* in the beam is zero along the neutral axis at its centre, and is a maximum at the surface, at the mid-point of the beam (because the

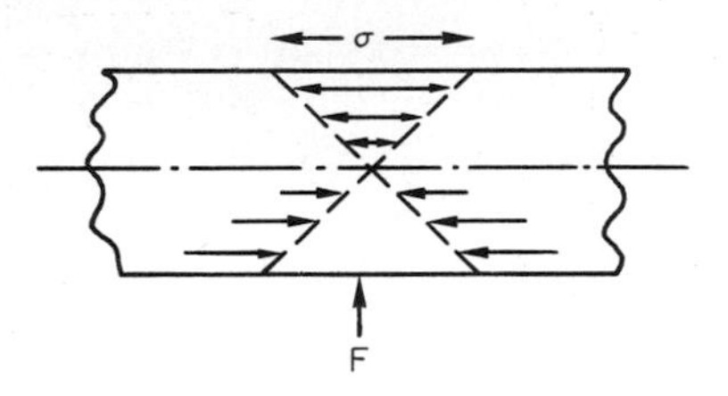

Fig. 12.2. Stresses inside a leaf spring.

bending moment is biggest there). The maximum surface stress is given by

$$\sigma = \frac{3Fl}{2bt^2}. \tag{12.2}$$

Now, to be successful, a spring must not undergo a permanent set during use: it must always "spring" back. The condition for this is that the maximum stress (eqn. (12.2)) always be less than the yield stress:

$$\frac{3Fl}{2bt^2} < \sigma_y. \tag{12.3}$$

Substituting this into eqn. (12.1) to eliminate F gives

$$\left(\frac{\sigma_y}{E}\right) > \frac{6\delta t}{l^2}. \tag{12.4}$$

This equation says: if in service a spring has to undergo a given deflection, δ, the ratio of σ_y/E must be high enough to avoid a permanent set. This is why we have listed values of σ_y/E in Table 12.1. The best springs are made of materials with high σ_y/E. For this reason spring materials are heavily strengthened (see Chapter 10): by solid solution strengthening plus work-hardening (cold-rolled, single-phase brass and bronze), solid solution and precipitate strengthening (spring steel), and so on. Annealing any spring material removes the work-hardening, and may cause the precipitate to coarsen (increasing the particle spacing), making the material useless as a spring.

Example: *Springs for a centrifugal clutch.* Suppose that you are asked to select a material for a spring with the following application in mind. A spring-controlled clutch like that shown in Fig. 12.3 is designed to transmit 20 horse power at 800 rpm; the clutch is to begin to pick up load at 600 rpm. The blocks are lined with Ferodo or some other friction material. When properly adjusted, the maximum deflection of the springs is to be 6.35 mm (but the friction pads may wear, and larger deflections may occur; this is a standard problem with springs—almost always, they must withstand occasional extra deflections without losing their sets).

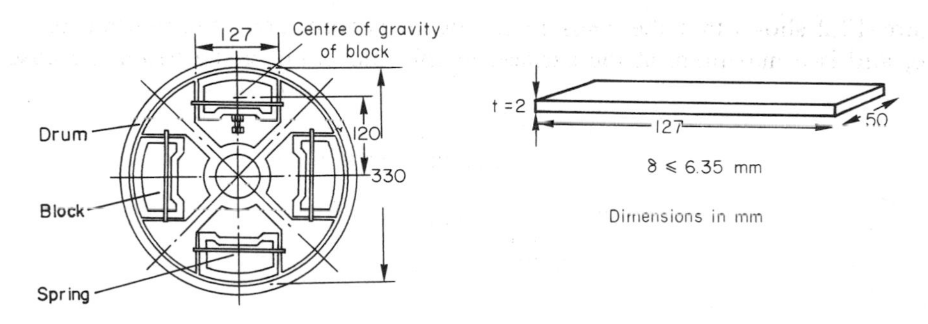

Fig. 12.3. Leaf springs in a centrifugal clutch.

Mechanics

The force on the spring is

$$F = Mr\omega^2 \tag{12.5}$$

where M is the mass of the block, r the distance of the centre of gravity of the block from the centre of rotation, and ω the angular velocity. The *net* force the block exerts on the clutch rim at full speed is

$$Mr(\omega_2^2 - \omega_1^2) \tag{12.6}$$

where ω_2 and ω_1 correspond to the angular velocities at 800 and 600 rpm (the *net* force must be zero for $\omega_2 = \omega_1$, at 600 rpm). The full power transmitted is given by

$$4\mu_s Mr(\omega_2^2 - \omega_1^2) \times \text{distance moved per second by inner rim of clutch at full speed,}$$

i.e.

$$\text{power} = 4\mu_s Mr(\omega_2^2 - \omega_1^2) \times \omega_2 r \tag{12.7}$$

where μ_s is the coefficient of static friction. r is specified by the design (the clutch cannot be too big) and μ_s is a constant (partly a property of the clutch-lining material). Both the power and ω_2 and ω_1 are specified in eqn. (12.7), so M is specified also; and finally the maximum force on the spring, too, is determined by the design from $F = Mr\omega_1^2$. The requirement that this force deflect the beam by only 6.35 mm with the linings just in contact is what determines the thickness, t, of the spring via eqn. (12.1) (l and b are fixed by the design).

Metallic materials for the clutch springs

Given the spring dimensions ($t = 2$ mm, $b = 50$ mm, $l = 127$ mm) and given $\delta \leq$ 6.35 mm, all specified by design, which material should we use? Equation (12.4) gives

$$\left(\frac{\sigma_y}{E}\right) > \frac{6 \times 6.35 \times 2}{127 \times 127} = 4.7 \times 10^{-3}. \tag{12.8}$$

Table 12.1 shows that spring steel, the cheapest material listed, is adequate for this purpose, but has a worryingly small safety factor to allow for wear of the linings. Only the expensive beryllium–copper alloy, of all the metals shown, would give a significant safety factor ($\sigma_y/E = 11.5 \times 10^{-3}$).

In many designs, the mechanical requirements are such that single springs of the type considered so far would yield even if made from beryllium copper. This commonly arises in the case of suspension springs for vehicles, etc., where both large δ ("soft" suspensions) and large F (good load-bearing capacity) are required. The solution then can be to use multi-leaf springs (Fig. 12.4). t can be made *small* to give *large* δ without yield according to

$$\left(\frac{\sigma_y}{E}\right) > \frac{6\delta t}{l^2}, \tag{12.9}$$

whilst the lost load-carrying capacity resulting from small t can be made up by having several leaves combining to support the load.

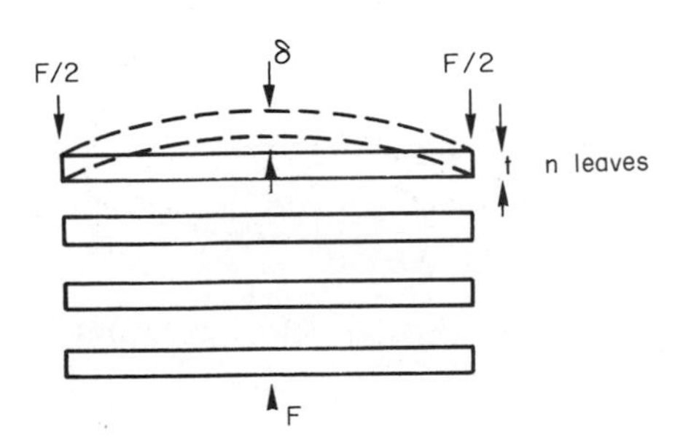

Fig. 12.4. Multi-leaved springs (schematic).

Non-metallic materials

Finally, materials other than the metals originally listed in Table 12.1 can make good springs. Glass, or fused silica, with σ_y/E as large as 58×10^{-3} is excellent, *provided* it operates under protected conditions where it cannot be scratched or suffer impact loading (it has long been used for galvanometer suspensions). Nylon is good—provided the forces are low—having $\sigma_y/E\approx22\times10^{-3}$, and it is widely used in household appliances and children's toys (you probably brushed your teeth with little nylon springs this morning). Leaf springs for heavy trucks are now being made of CFRP: the value of σ_y/E (6×10^{-3}) is similar to that of spring steel, and the weight saving compensates for the higher cost. CFRP is always worth examining where an innovative use of materials might offer advantages.

Case study 2: plastic design: materials for a pressure vessel

We shall now examine material selection for a pressure vessel able to contain a gas at pressure p, first minimising the *weight*, and then the *cost*. We shall seek a design that will not fail by plastic collapse (i.e. general yield). But we must be cautious: structures can also fail by *fast fracture*, by *fatigue*, and by *corrosion* superimposed on these other modes of failure. We shall discuss these in Chapters 13, 15 and 23. Here we shall assume that plastic collapse is our only problem.

Pressure vessel of minimum weight

The body of an aircraft, the hull of a spacecraft, the fuel tank of a rocket: these are examples of pressure vessels which must be as light as possible.

The stress in the vessel wall (Fig. 12.5) is:

$$\sigma = \frac{pr}{2t}. \tag{12.10}$$

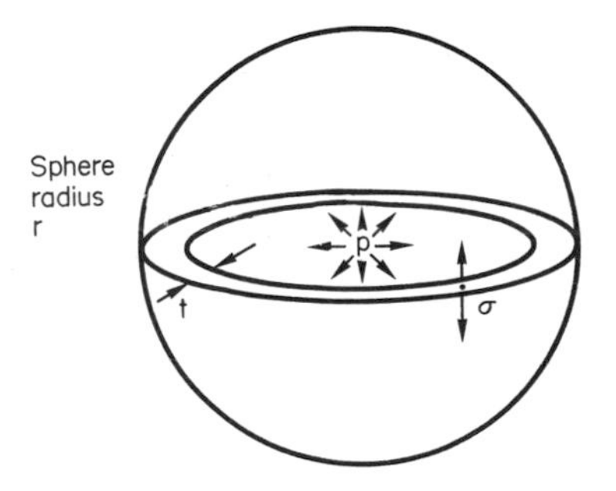

Fig. 12.5. Thin-walled spherical pressure vessel.

r, the radius of the pressure vessel, is fixed by the design. For safety, $\sigma \le \sigma_y/S$, where S is the safety factor. The vessel mass is

$$M = 4\pi r^2 t\rho \tag{12.11}$$

so that

$$t = \frac{M}{4\pi r^2 \rho}. \tag{12.12}$$

Substituting for t in eqn. (12.8) we find that

$$\frac{\sigma_y}{S} \ge \frac{pr}{2}\frac{4\pi r^2 \rho}{M} = \frac{2\pi p r^3 \rho}{M}. \tag{12.13}$$

From eqn. (12.11) we have, for the mass,

$$M = S 2\pi p r^3 \left(\frac{\rho}{\sigma_y}\right) \tag{12.14}$$

so that for the lightest vessel we require the smallest value of (ρ/σ_y). Table 12.2 gives values of ρ/σ_y for candidate materials.

TABLE 12.2
MATERIALS FOR PRESSURE VESSELS

Material	σ_y/MN m^{-2}	ρ/Mg m^{-3}	$\tilde{p}$/UK£(US$) tonne^{-1}	$\frac{\rho}{\sigma_y} \times 10^6$/ s^2 m^{-2}	$\frac{\tilde{p}\rho}{\sigma_y} \times 10^6$/ UK£(US$) m^{-1} N^{-1}
Reinforced concrete	200	2.5	130 (290)	13	1.6 (3.5)
Alloy steel (pressure-vessel steel)	1000	7.8	500 (1100)	7.8	3.9 (8.6)
Mild steel	220	7.8	220 (490)	36	7.9 (17)
Aluminium alloy	400	2.7	1,000 (2200)	6.8	6.8 (15)
Fibreglass	200	1.8	1,100 (2420)	9.0	9.9 (22)
CFRP	600	1.5	90,000 (198,000)	2.5	230 (510)

By far the lightest pressure vessel is that made of CFRP. Aluminium alloy and pressure-vessel steel come next. Reinforced concrete or mild steel results in a very heavy vessel.

Pressure vessel of minimum cost

If the cost of the material is $\tilde{p}$ UK£(US$) tonne^{-1} then the material cost of the vessel is

$$\tilde{p}M = \text{constant } \tilde{p}\left(\frac{\rho}{\sigma_y}\right). \tag{12.15}$$

Thus material costs are minimised by minimising $\tilde{p}(\rho/\sigma_y)$. Data are given in Table 12.2.

The proper choice of material is now a quite different one. Reinforced concrete is now the best choice—that is why many water towers, and pressure vessels for nuclear reactors, are made of reinforced concrete. After that comes pressure-vessel steel—it offers the best compromise of both price and weight. GFRP and CFRP are very expensive.

Case study 3: large-strain plasticity—rolling of metals

Forging, *sheet drawing* and *rolling* are metal-forming processes by which the section of a billet or slab is reduced by compressive plastic deformation. When a slab is rolled (Fig. 12.6) the section is reduced from t_1 to t_2 over a length l as it passes through the rolls. At first sight, it might appear that there would be no sliding (and thus no friction) between the slab and the rolls, since these move with the slab. But the metal is elongated in the rolling direction, so it speeds up as it passes through the rolls, and some slipping is inevitable. If the rolls are polished and lubricated (as they are for precision and cold-rolling) the frictional losses are small. We shall ignore them here (though all detailed treatments of rolling include them) and calculate the *rolling torque* for perfectly lubricated rolls.

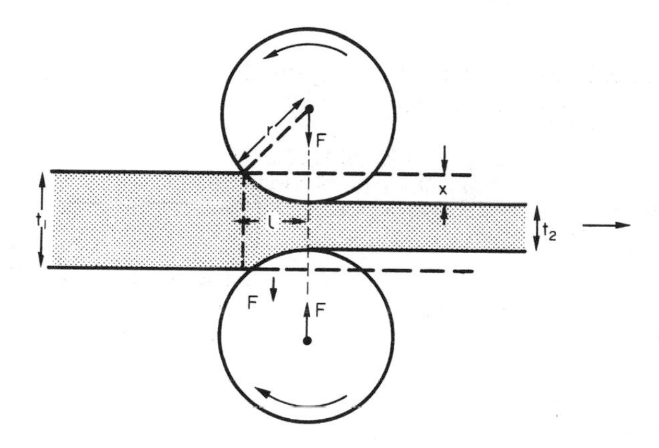

Fig. 12.6. The rolling of metal sheet.

From the geometry of Fig. 12.6

$$l^2 + (r - x)^2 = r^2$$

or, if $x = \frac{1}{2}(t_1 - t_2)$ is small (as it almost always is),

$$l = \sqrt{r(t_1 - t_2)}.$$

The rolling force F must cause the metal to yield over the length l and width w (normal to Fig. 12.6). Thus

$$F = \sigma_y wl.$$

If the reaction on the rolls appears halfway along the length marked l, as shown on the lower roll, the torque is

$$T = \frac{Fl}{2}$$

$$= \frac{\sigma_y wl^2}{2},$$

giving

$$T = \frac{\sigma_y w r (t_1 - t_2)}{2}. \tag{12.16}$$

The torque required to drive the rolls increases with yield strength σ_y, so hot-rolling (when σ_y is low—see Chapter 17) takes less power than cold-rolling. It obviously increases with the reduction in section $(t_1 - t_2)$. And it increases with roll diameter $2r$; this is one of the reasons why small-diameter rolls, often backed by two or more rolls of larger diameter (simply to stop them bending), are used.

Rolling can be analysed in much more detail to include important aspects which we have ignored: friction, the elastic deformation of the rolls, and the constraint of plane strain imposed by the rolling geometry. But this case study gives an idea of why an understanding of plasticity, and the yield strength, is important in forming operations, both for metals and polymers.

Further reading

C. R. Calladine, *Engineering Plasticity*, Pergamon Press, 1969.
R. Hill, *The Mathematical Theory of Plasticity*, Oxford University Press, 1950.
W. A. Backofen, *Deformation Processing*, Addison-Wesley, 1972.

D. Fast fracture, toughness and fatigue

CHAPTER 13

FAST FRACTURE AND TOUGHNESS

Introduction

Sometimes, structures which were properly designed to avoid both excessive elastic deflection and plastic yielding fail in a catastrophic way by *fast fracture*. Common to these failures—of things like welded ships, welded bridges and gas pipelines and pressure vessels under large internal pressures—is the presence of cracks, often the result of imperfect welding. Fast fracture is caused by the growth—at the speed of sound in the material—of existing cracks that suddenly became unstable. Why do they do this?

Energy criterion for fast fracture

If you blow up a balloon, energy is stored in it. There is the energy of the compressed gas in the balloon, and there is the elastic energy stored in the rubber membrane itself. As you increase the pressure, the total amount of elastic energy in the system increases.

If we then introduce a flaw into the system, by poking a pin into the inflated balloon, the balloon will explode, and all this energy will be released. The membrane fails by fast fracture, *even though well below its yield strength*. But if we introduce a flaw of the same dimensions into a system with *less* energy in it, as when we poke our pin into a *partially* inflated balloon, the flaw is stable and fast fracture does not occur. Finally, if we blow up the punctured balloon progressively, we eventually reach a pressure at which it suddenly bursts. In other words, we have arrived at a *critical* balloon *pressure* at which our pin-sized flaw is just unstable, and fast fracture *just* occurs. Why is this?

To make the flaw grow, say by 1 mm, we have to tear the rubber to create 1 mm of new crack surface, and this consumes energy: the tear energy of the rubber per unit area × the area of surface torn. If the work done by the gas pressure inside the balloon, plus the release of elastic energy from the membrane itself, is less than this energy the tearing simply cannot take place—it would infringe the laws of thermodynamics.

We can, of course, increase the energy in the system by blowing the balloon up a bit more. The crack or flaw will remain stable (i.e. it will not grow) until the system (balloon plus compressed gas) has stored in it enough energy that, if the crack advances, *more energy is released than is absorbed.* There is, then, a *critical pressure* for fast fracture of a pressure vessel containing a crack or flaw of a given *size*.

All sorts of accidents (the sudden collapsing of bridges, sudden explosion of steam boilers) have occurred—and still do—due to this effect. In all cases, the critical stress—above which enough energy is available to provide the tearing energy needed to make the crack advance—was exceeded, taking the designer completely by surprise. But how do we calculate this critical stress?

From what we have said already, we can write down an energy balance which must be met if the crack is to advance, and fast fracture is to occur. Suppose a crack of length a in a material of thickness t advances by δa, then we require that: work done by loads $\geq$ change of elastic energy + energy absorbed at the crack tip, i.e.

$$\delta W \geq \delta U^{el} + G_c t \delta a \tag{13.1}$$

where G_c is the energy absorbed per unit area of *crack* (*not* unit area of new surface), and $t\delta a$ is the crack area.

G_c is a material property—it is the energy absorbed in making unit area of crack, and we call it the *toughness* (or, sometimes, the "critical strain energy release rate"). Its units are energy m^{-2} or $J\,m^{-2}$. A high toughness means that it is hard to make a crack propagate (as in copper, for which $G_c \approx 10^6\,J\,m^{-2}$). Glass, on the other hand, cracks very easily; G_c for glass is only $\approx 10\,J\,m^{-2}$.

This same quantity G_c measures the strength of adhesives. You can measure it for the adhesive used on sticky tape (like Sellotape) by hanging a weight on a partly peeled length while supporting the roll so that it can freely rotate (hang it on a pencil) as shown in Fig. 13.1. Increase the load to the value M that just causes rapid peeling

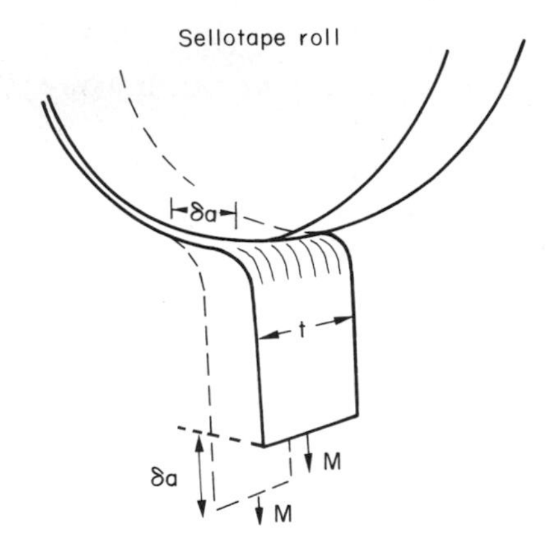

Fig. 13.1. How to determine G_c for Sellotape adhesive.

(= fast fracture). For this geometry, the quantity δU^{el} is small compared to the work done by M (the tape has comparatively little "give") and it can be neglected. Then, from our energy formula,

$$\delta W = G_c t \delta a$$

for fast fracture. In our case,

$$Mg\delta a = G_c t \delta a,$$

$$Mg = G_c t,$$

and therefore,

$$G_c = \frac{Mg}{t}.$$

Typically, $t = 2$ cm, $M = 1$ kg and $g \approx 10$ m s^{-2}, giving

$$G_c \approx 500 \text{ J m}^{-2}.$$

This is a reasonable value for adhesives, and a value bracketed by the values of G_c for many polymers.

Naturally, in most cases, we cannot neglect δU^{el}, and must derive more general relationships. Let us first consider a cracked plate of material loaded so that the displacements at the boundary of the plate are fixed. This is a common mode of loading a material—it occurs frequently in welds between large pieces of steel, for example—and is one which allows us to calculate δU^{el} quite easily.

Fast fracture at fixed displacements

The plate shown in Fig. 13.2 is clamped under tension so that its upper and lower ends are fixed. Since the ends cannot move, the forces acting on them can do no work,

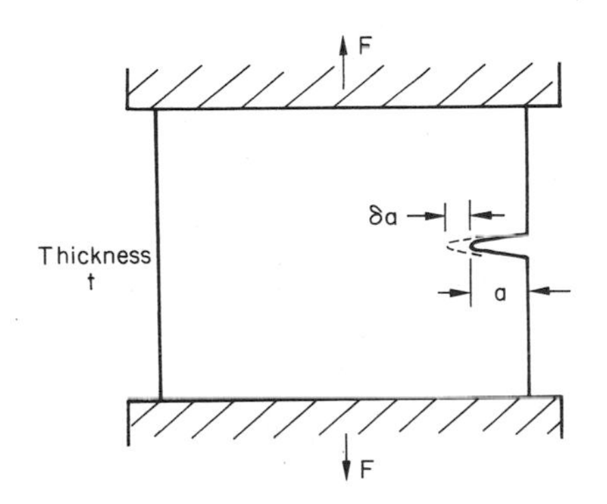

Fig. 13.2. Fast fracture in a fixed plate.

and $\delta W = 0$. Accordingly, our energy formula gives, for the onset of fast fracture,

$$-\delta U^{el} = G_c t \delta a. \qquad (13.2)$$

Now, as the crack grows into the plate, it allows the material of the plate near the crack to *relax*, so that it becomes less highly stressed, and *loses* elastic energy. δU^{el} is thus *negative*, so that $-\delta U^{el}$ is *positive*, as it must be since G_c is defined positive. We can estimate δU^{el} in the way shown in Fig. 13.3.

Let us examine a small cube of material of unit volume inside our plate. Due to the load F this cube is subjected to a stress σ, producing a strain ε. Each unit cube therefore has strain energy U^{el} of $\frac{1}{2}\sigma\varepsilon$, or $\sigma^2/2E$. If we now introduce a crack of length a,

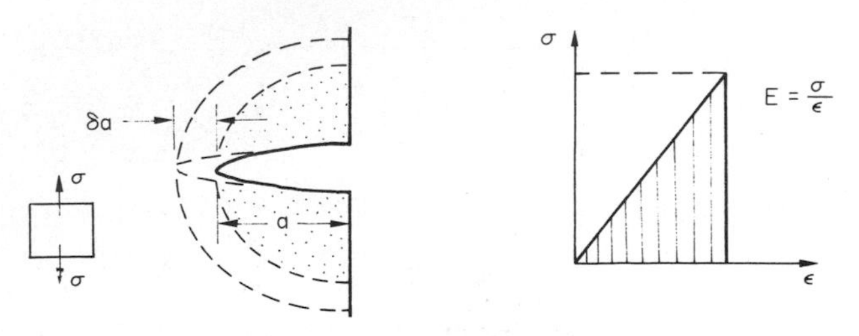

Fig. 13.3. The release of stored strain energy as a crack grows.

we can consider that the material in the dotted region relaxes (to zero stress) so as to lose all its strain energy. The energy change is

$$U^{el} = -\frac{\sigma^2}{2E}\frac{\pi a^2 t}{2}.$$

As the crack spreads by length δa, we can calculate the appropriate δU^{el} as

$$\delta U^{el} = \frac{dU^{el}}{da}\delta a = -\frac{\sigma^2}{2E}\frac{2\pi a t}{2}\delta a.$$

The critical condition (eqn. (13.2)) then gives

$$\frac{\sigma^2 \pi a}{2E} = G_c$$

at onset of fast fracture.

Actually, our assumption about the way in which the plate material relaxes is obviously rather crude, and a rigorous mathematical solution of the elastic stresses and strains around the crack indicates that our estimate of δU^{el} is too low by exactly a factor of 2. Thus, correctly, we have

$$\frac{\sigma^2 \pi a}{E} = G_c,$$

which reduces to

$$\sigma\sqrt{\pi a} = \sqrt{EG_c} \qquad (13.3)$$

at fast fracture.

Fast fracture at fixed loads

Another, obviously very common way of loading a plate of material, or any other component for that matter, is simply to hang weights on it (fixed loads) (Fig. 13.4). Here the situation is a little more complicated than it was in the case of fixed displacements. As the crack grows, the plate becomes less *stiff*, and relaxes so that the applied forces move and do work. δW is therefore finite and positive. However, δU^{el} is now positive also (it turns out that some of δW goes into increasing the strain energy of the plate) and our final result for fast fracture is in fact found to be unchanged.

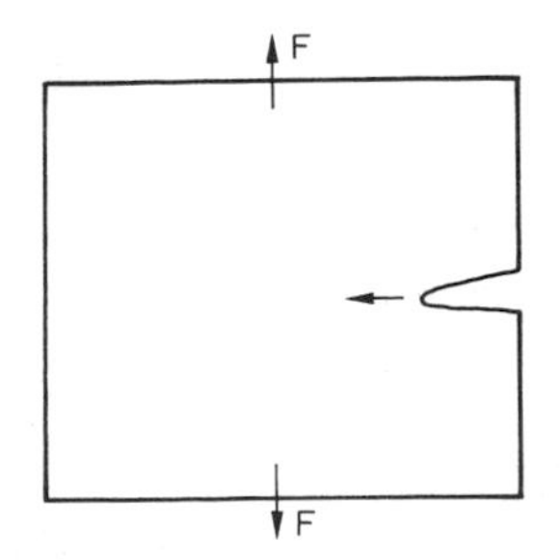

Fig. 13.4. Fast fracture of a dead-loaded plate.

The fast-fracture condition

Let us now return to our condition for the onset of fast fracture, knowing it to be general* for engineering structures

$$\sigma\sqrt{\pi a} = \sqrt{EG_c}.$$

The left-hand side of our equation says that *fast fracture will occur when, in a material subjected to a stress* σ, *a crack reaches some critical size* a: or, alternatively, when *material containing cracks of size* a *is subjected to some critical stress* σ. The right-hand side of our result *depends on material properties only*; E is obviously a material constant, and G_c, the energy required to generate unit area of crack, again must depend only on the basic properties of our material. Thus, the important point about the equation is that *the critical combination of stress and crack length at which fast fracture commences is a material constant.*

The term $\sigma\sqrt{\pi a}$ crops up so frequently in discussing fast fracture that it is usually abbreviated to a single symbol, K, having units $MN\,m^{-3/2}$; it is called, somewhat unclearly, the *stress intensity factor.* Fast fracture therefore occurs when

$$K = K_c$$

where K_c $(=\sqrt{EG_c})$ is the *critical* stress intensity factor, more usually called the *fracture toughness.*

To summarise:

G_c = *toughness* (sometimes, critical strain energy release rate). Usual units: $kJ\,m^{-2}$;
$K_c = \sqrt{EG_c}$ = *fracture toughness* (sometimes: critical stress intensity factor). Usual units: $MN\,m^{-3/2}$;
$K = \sigma\sqrt{\pi a}$ = stress intensity factor*. Usual units: $MN\,m^{-3/2}$.

Fast fracture occurs when $K = K_c$.

Data for G_c and K_c

K_c can be determined experimentally for any material by inserting a crack of known length a into a piece of the material and loading until fast fracture occurs. G_c can be

* But see note at end of this chapter.

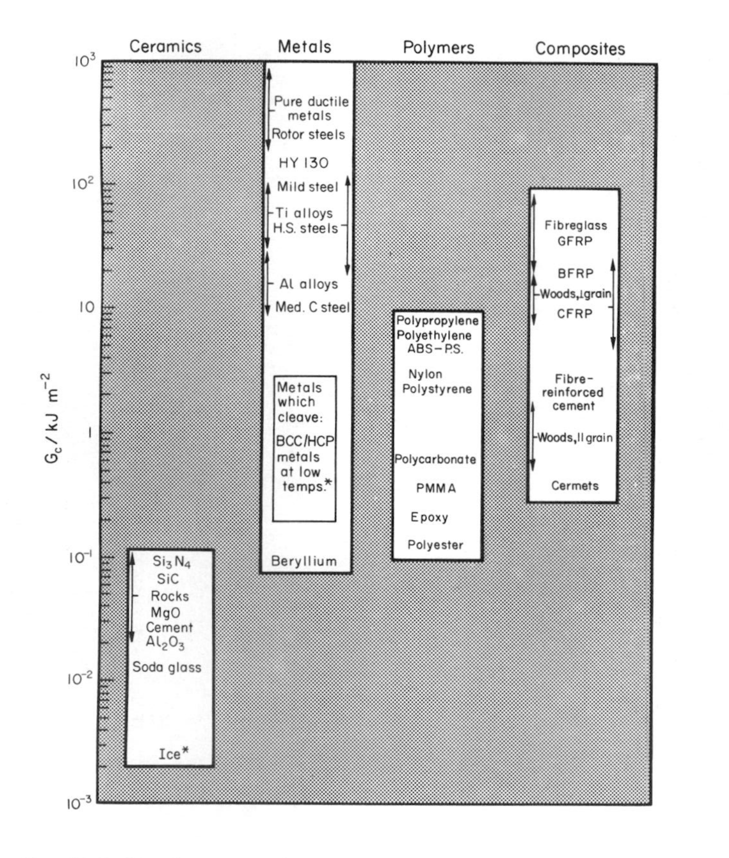

Fig. 13.5. Toughness, G_c (values at room temperature unless starred).

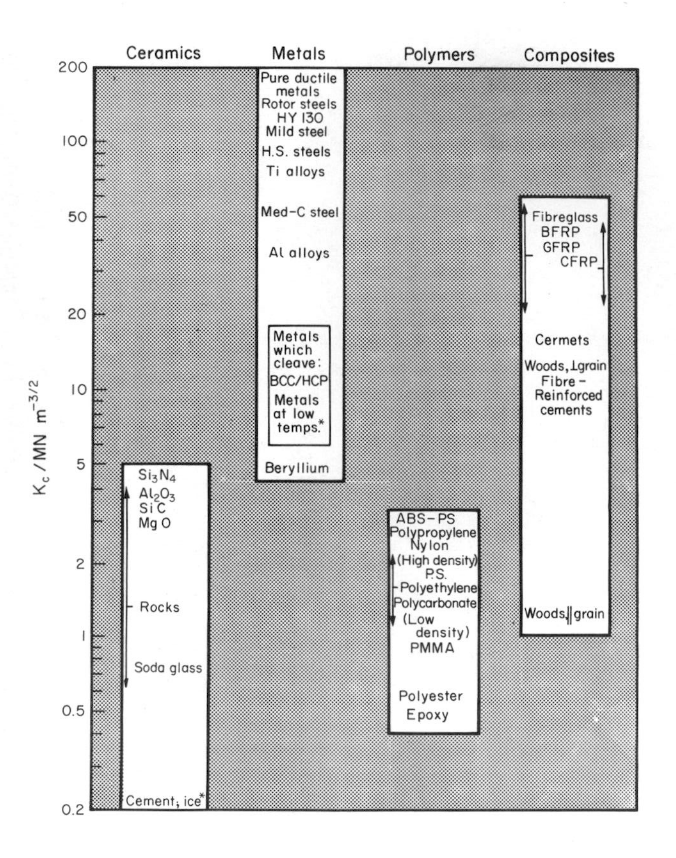

Fig. 13.6. Fracture toughness, K_c (values at room temperature unless starred).

TABLE 13.1

TOUGHNESS, G_c, AND FRACTURE TOUGHNESS, K_c

Material	G_c/kJ m^{-2}	K_c/MN m$^{-3/2}$
Pure ductile metals (e.g. Cu, Ni, Ag, Al)	100–1000	100–350
Rotor steels (A533; Discalloy)	220–240	204–214
Pressure-vessel steels (HY130)	150	170
High-strength steels (HSS)	15–118	50–154
Mild steel	100	140
Titanium alloys (Ti 6Al 4V)	26–114	55–115
GFRPs	10–100	20–60
Fibreglass (glassfibre epoxy)	40–100	42–60
Aluminium alloys (high strength–low strength)	8–30	23–45
CFRPs	5–30	32–45
Common woods, crack ⊥ to grain	8–20	11–13
Boron-fibre expoxy	17	46
Medium-carbon steel	13	51
Polypropylene	8	3
Polyethylene (low density)	6–7	1
Polyethylene (high density)	6–7	2
ABS polystyrene	5	4
Nylon	2–4	3
Steel-reinforced cement	0.2–4	10–15
Cast iron	0.2–3	6–20
Polystyrene	2	2
Common woods, crack ‖ to grain	0.5–2	0.5–1
Polycarbonate	0.4–1	1.0–2.6
Cobalt/tungsten carbide cermets	0.3–0.5	14–16
PMMA	0.3–0.4	0.9–1.4
Epoxy	0.1–0.3	0.3–0.5
Granite (Westerly Granite)	0.1	3
Polyester	0.1	0.5
Silicon nitride, Si_3N_4	0.1	4–5
Beryllium	0.08	4
Silicon carbide SiC	0.05	3
Magnesia, MgO	0.04	3
Cement/concrete, unreinforced	0.03	0.2
Calcite (marble, limestone)	0.02	0.9
Alumina, Al_2O_3	0.02	3–5
Shale (oilshale)	0.02	0.6
Soda glass	0.01	0.7–0.8
Electrical porcelain	0.01	1
Ice	0.003	0.2*

* Values at room temperature unless starred.

derived from the data for K_c using the relation $K_c = \sqrt{EG_c}$. Figures 13.5 and 13.6 and Table 13.1 show experimental data for K_c and G_c for a wide range of metals, polymers, ceramics and composites. The values of K_c and G_c range considerably, from the least tough materials, like ice and ceramics, to the toughest, like ductile metals; polymers have intermediate toughness, G_c, but low fracture toughness, K_c (because their *moduli* are low). However, reinforcing polymers to make *composites* produces materials having good fracture toughnesses. Finally, although most metals are tough at or above room temperature, when many (e.g. b.c.c. metals like steels, or h.c.p. metals) are cooled sufficiently, they become quite brittle as the data show.

Obviously these figures for toughness and fracture toughness are extremely important—ignorance of such data has led, and can continue to lead, to engineering disasters of the sort we mentioned at the beginning of this chapter. But just how do these large variations between various materials arise? Why *is* glass so brittle and annealed copper so tough? We shall explain why in Chapter 14.

A note on the Stress Intensity, K

On pp. 124 and 125 we showed that

$$K = \sigma\sqrt{\pi a} = \sqrt{EG_c}$$

at onset of fast fracture. Strictly speaking, this result is valid only for a *thin, semi-infinite* plate of material such that, on p. 123 $t \ll a \ll$ other plate dimensions. In practice, of course, the problems we encounter seldom satisfy this condition, and a numerical correction to $\sigma\sqrt{\pi a}$ is required to get the strain energy calculation right. For example, when t and other plate dimensions $\gg a$, we have

$$K = 1.12\sigma\sqrt{\pi a}.$$

Other geometries (e.g. plate with $t \gg a$ but other dimensions $\approx a$) have other correction factors. In this book we have ignored these corrections. They tend at the present level to clutter up the basic physics of crack propagation, which remains unchanged; and often the factors are close enough to unity to be glossed over in any case. But if you find yourself in the position of doing *detailed* design calculations on fast fracture, *do* remember these correction factors—their values can easily be found from tables in standard reference books such as that listed under **Further reading**.

Further reading

R. W. Hertzberg, *Deformation and Fracture Mechanics of Engineering Materials*, Wiley, 1976, Chap. 8.
B. R. Lawn and T. R. Wilshaw, *Fracture of Brittle Solids*, Cambridge University Press, 1975, Chap. 3.
J. F. Knott, *Fundamentals of Fracture Mechanics*, Butterworths, 1973, Chap. 4.
H. Tada, P. Paris and G. Irwin, *The Stress Analysis of Cracks Handbook*, Del Research Corporation, St Louis, 1973 (for Tabulation of Stress Intensities).

CHAPTER 14

MICROMECHANISMS OF FAST FRACTURE

In Chapter 13 we showed that, if a material contains a crack, and is sufficiently stressed, the crack becomes unstable and grows—at up to the speed of sound in the material—to cause catastrophically rapid fracture, or *fast fracture* at a stress less than the yield stress. We were able to quantify this phenomenon and obtained a relationship for the onset of fast fracture

$$\sigma\sqrt{\pi a} = \sqrt{EG_c}$$

or, in more succinct notation,

$$K = K_c \quad \text{for fast fracture.}$$

It is helpful to compare this with other, similar, "failure" criteria:

$$\sigma = \sigma_y \quad \text{for yielding,}$$
$$M = M_p \quad \text{for plastic collapse,}$$
$$P/A = H \quad \text{for indentation.}$$

(Here M is the moment and M_p the fully-plastic moment of, for instance, a beam; P/A is the indentation pressure and H the hardness of, for example, armour plating.) The left-hand side of each of these equations describes the *loading conditions*; the right-hand side is a *material property*. When the left-hand side (which increases with load) equals the right-hand side (which is fixed), failure occurs.

Some materials, like glass, have low G_c and K_c, and crack easily; ductile metals have high G_c and K_c and are very resistant to fast-fracture; polymers have intermediate G_c, but can be made tougher by making them into composites; and (finally) many metals, when cold, become brittle—that is, G_c and K_c fall with temperature. How can we explain these important observations?

Mechanisms of crack propagation, 1: ductile tearing

Let us first of all look at what happens when we load a cracked piece of a *ductile* metal—in other words, a metal that can flow readily to give large plastic deformations (like pure copper; or mild steel at, or above, room temperature). If we load the material sufficiently, we can get fracture to take place starting from the crack. If you

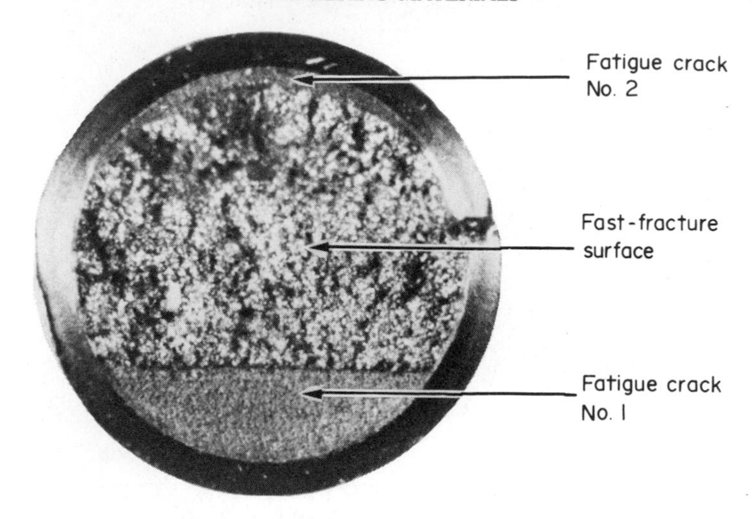

Fig. 14.1. Before it broke, this steel bolt held a seat onto its mounting at Milan airport. Whenever someone sat down, the lower part of the cross-section went into tension, causing a crack to grow there by *metal fatigue* (Chapter 15; crack No. 1). When someone got up again, the upper part went into tension, causing fatigue crack No. 2 to grow. Eventually the bolt failed by fast fracture from the larger of the two fatigue cracks. The victim was able to escape with the fractured bolt!

examine the surfaces of the metal after it has fractured (Fig. 14.1) you see that the fracture surface is extremely rough, indicating that a great deal of plastic work has taken place. Let us explain this observation. Whenever a crack is present in a material, the stress close to the crack, σ_{local}, is greater than the average stress σ applied to the piece of material; the crack has the effect of *concentrating* the stress. Mathematical analysis shows that the local stress ahead of a *sharp* crack in an elastic material is

$$\sigma_{\text{local}} = \sigma + \sigma\sqrt{\frac{a}{2r}}. \tag{14.1}$$

The closer one approaches to the tip of the crack, the higher the local stress becomes, until at some distance r_y from the tip of the crack the stress reaches the yield stress, σ_y, of the material, and plastic flow occurs (Fig. 14.2). The distance r_y is easily calculated by setting $\sigma_{\text{local}} = \sigma_y$ in eqn. (14.1). Assuming r_y to be small compared to the crack length, a, the result is

$$r_y = \frac{\sigma^2 a}{2\sigma_y^2} = \frac{K^2}{2\pi\sigma_y^2}. \tag{14.2}$$

The crack propagates when K is equal to K_c; the width of the *plastic zone*, r_y, is then given by eqn. (14.2) with K replaced by K_c. Note that the zone of plasticity shrinks rapidly as σ_y increases: cracks in soft metals have a large plastic zone; cracks in hard ceramics have a small zone, or none at all.

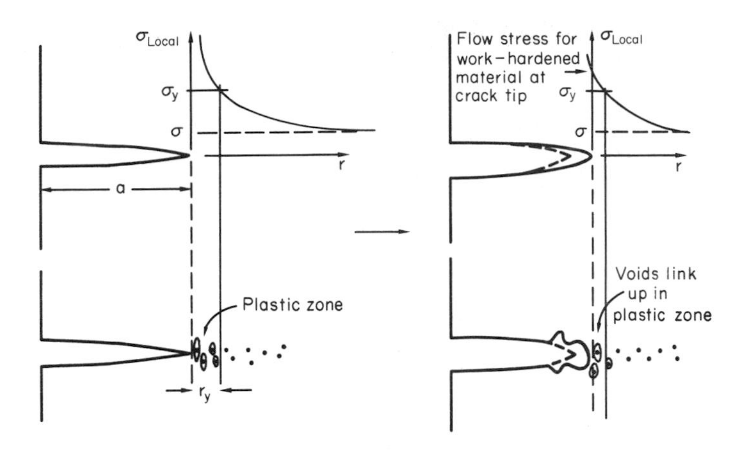

Fig. 14.2. Crack propagation by ductile tearing.

Even when nominally pure, most metals contain tiny *inclusions* (or particles) of chemical compounds formed by reaction between the metal and impurity atoms. Within the plastic zone, plastic flow takes place around these inclusions, leading to elongated cavities, as shown in Fig. 14.2. As plastic flow progresses, these cavities link up, and the crack advances by means of this *ductile tearing*. The plastic flow at the crack tip naturally turns our initially sharp crack into a *blunt* crack, and it turns out from the stress mathematics that this *crack blunting* decreases σ_{local} so that, at the crack tip itself, σ_{local} is just sufficient to keep on plastically deforming the work-hardened material there, as the diagram shows.

The important thing about crack growth by ductile tearing is that *it consumes a lot of energy by plastic flow*; the bigger the plastic zone, the more energy is absorbed. High energy absorption means that G_c is high, and so is K_c. This is why ductile metals are so tough. Other materials, too, owe their toughness to this behaviour—plasticine is one, and some polymers also exhibit toughening by processes similar to ductile tearing.

Mechanisms of crack propagation, 2: cleavage

If you now examine the fracture surface of something like a ceramic, or a glass, you see a very different state of affairs. Instead of a very rough surface, indicating massive local plastic deformation, you see a rather featureless, flat surface suggesting little or no plastic deformation. How is it that cracks in ceramics or glasses can spread without plastic flow taking place? Well, the local stress ahead of the crack tip, given by our formula

$$\sigma_{\text{local}} = \sigma + \sigma\sqrt{\frac{a}{2r}},$$

can clearly approach very high values very near to the crack tip *provided that blunting of our sharp crack tip does not occur.* As we showed in Chapter 8, ceramics and glasses

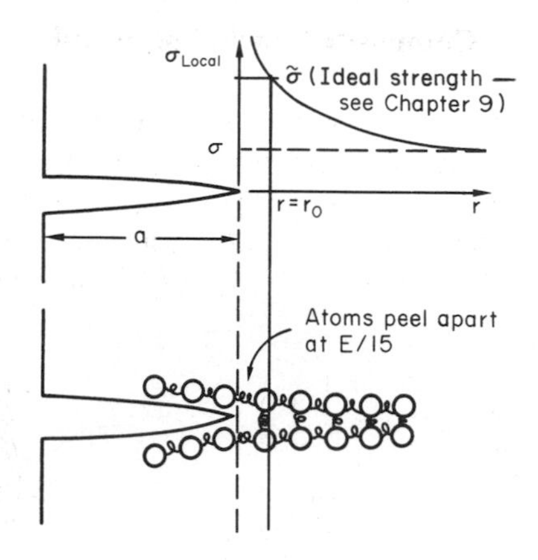

Fig. 14.3. Crack propagation by cleavage.

have very high yield strengths, and thus very little plastic deformation takes place at crack tips in these materials. Even allowing for a small degree of crack blunting, the local stress at the crack tip is still in excess of the ideal strength and is thus large enough to literally break apart the interatomic bonds there; the crack then spreads between a pair of atomic planes giving rise to an atomically flat surface by *cleavage*. The energy required simply to break the interatomic bonds is *much* less than that absorbed by ductile tearing in a tough material, and this is why materials like ceramics and glasses are so brittle. It is also why some steels become brittle and fail like glass, at low temperatures—as we shall now explain.

At low temperatures metals having b.c.c. and h.c.p. structures become brittle and fail by cleavage, even though they may be tough at or above room temperature. In fact, only those metals with an f.c.c. structure (like copper, lead, aluminium) remain unaffected by temperature in this way. In metals not having an f.c.c. structure, the motion of dislocations is assisted by the *thermal agitation* of the atoms (we shall talk in more detail about *thermally activated* processes in Chapter 18). At lower temperatures this thermal agitation is less, and the dislocations cannot move as easily as they can at room temperature in response to a stress—the intrinsic lattice resistance (Chapter 10) increases. The result is that the yield strength rises, and the plastic zone at the crack tip shrinks until it becomes so small that the fracture mechanism changes from ductile tearing to cleavage. This effect is called the *ductile-to-brittle* transition; for steels it can be as high as ≈0°C, depending on the composition of the steel; steel structures like ships, bridges and oil rigs are much more likely to fail in winter than in summer.

A somewhat similar thing happens in many polymers at the *glass–rubber transition* that we mentioned in Chapter 6. Below the transition these polymers are much more brittle than above it, as you can easily demonstrate by cooling a piece of rubber or polyethylene in liquid nitrogen. (Many other polymers, like epoxy resins, have low G_c values at *all* temperatures simply because they are heavily cross-linked at all temperatures by *covalent* bonds and the material does not flow at the crack tip to cause blunting.)

Composites, including wood

As Figs. 13.5 and 13.6 show, composites are tougher than ordinary polymers. The low toughness of materials like epoxy resins, or polyester resins, can be enormously increased by reinforcing them with carbon fibre or glass fibre. But why is it that putting a second, equally (or more) brittle material like graphite or glass into a brittle polymer makes a tough composite? The reason is that the *fibres act as crack stoppers* (Fig. 14.4).

Fig. 14.4. Crack stopping in composites.

The sequence in the diagram shows what happens when a crack runs through the brittle matrix towards a fibre. As the crack reaches the fibre, the stress field just ahead of the crack separates the matrix from the fibre over a small region (a process called *debonding*) and the crack is blunted so much that its motion is *arrested*. Naturally, this only works if the crack is running normal to the fibres: wood is very tough across the grain, but can be split easily (meaning that G_c is low) along it. One of the reasons why fibre composites are so useful in engineering design—in addition to their high *stiffnesses* that we talked about in Chapter 6—is their high *toughness* produced in this way. Of course, there are other ways of making polymers tough. The addition of small particles ("fillers") of various sorts to polymers can modify their properties considerably. Rubber-toughened polymers (like ABS), for example, derive their toughness from the small rubber particles they contain. A crack intersects and stretches them as shown in Fig. 14.5. The particles act as little springs, clamping the crack shut, and thereby increasing the load needed to make it propagate.

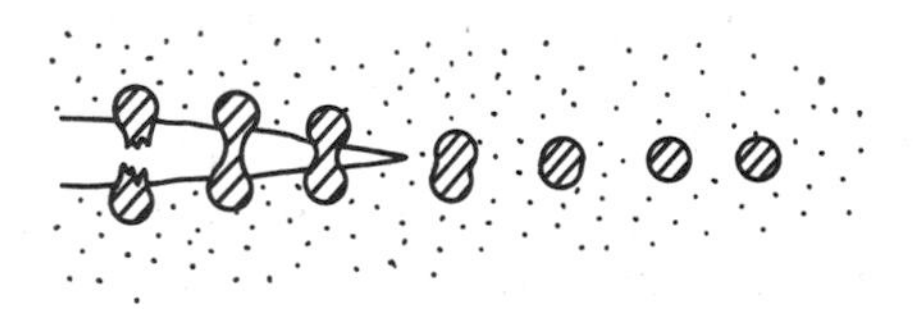

Fig. 14.5. Rubber-toughened polymers.

Avoiding brittle alloys

Let us finally return to the toughnesses of metals and alloys, as these are by far the most important class of materials for highly stressed applications. Even at, or above, room temperature, when nearly all common pure metals are tough, alloying of these metals with other metals or elements (e.g. with carbon to produce steels) can reduce the toughness. This is because alloying increases the resistance to dislocation motion (Chapter 10), raising the yield strength and causing the plastic zone to shrink. A more marked decrease in toughness can occur if enough impurities are added to make precipitates of chemical *compounds* formed between the metal and the impurities. These compounds can often be very brittle and, if they are present in the shape of extended plates (e.g. sigma-phase in stainless steel; graphite in cast iron), cracks can spread along the plates, leading to brittle fracture. Finally, heat treatments of alloys like steels can produce different *crystal structures* having great hardness (but also therefore great brittleness because crack blunting cannot occur). A good example of such a material is high-carbon steel after quenching into water from bright red heat: it becomes as brittle as glass. Proper heat treatment, following suppliers' specifications, is essential if materials are to have the properties you want. You will see an example of the unexpected results of faulty heat treatment in a Case Study given in Chapter 16.

Further reading

B. R. Lawn and T. R. Wilshaw, *Fracture of Brittle Solids*, Cambridge University Press, 1975, Chaps. 6 and 7.
J. F. Knott, *Fundamentals of Fracture Mechanics*, Butterworths, 1973, Chap. 8.

CHAPTER 15

FATIGUE FAILURE

Introduction

In the last two chapters we examined the conditions under which a crack was stable, and would not grow, and the condition

$$K = K_c$$

under which it would propagate catastrophically by fast fracture. If we know the maximum size of crack in the structure we can then choose a working load at which fast fracture will not occur.

But cracks can form, and grow slowly, at loads lower than this, if either the stress is cycled or if the environment surrounding the structure is corrosive (most are). The first process of slow crack growth—*fatigue*—is the subject of this lecture. The second—*corrosion*—is discussed later, in Chapters 21 to 24.

More formally: *if a component or structure is subjected to repeated stress cycles*, like the loading on the connecting rod of a petrol engine or on the wings of an aircraft—*it may fail at stresses well below the tensile strength, σ_{TS}, and often below the yield strength, σ_y,* of the material. The processes leading to this failure are termed "fatigue". When the clip of your pen breaks, when the pedals fall off your bicycle, when the handle of the refrigerator comes away in your hand, it is usually fatigue which is responsible.

We distinguish three categories of fatigue (Table 15.1).

TABLE 15.1

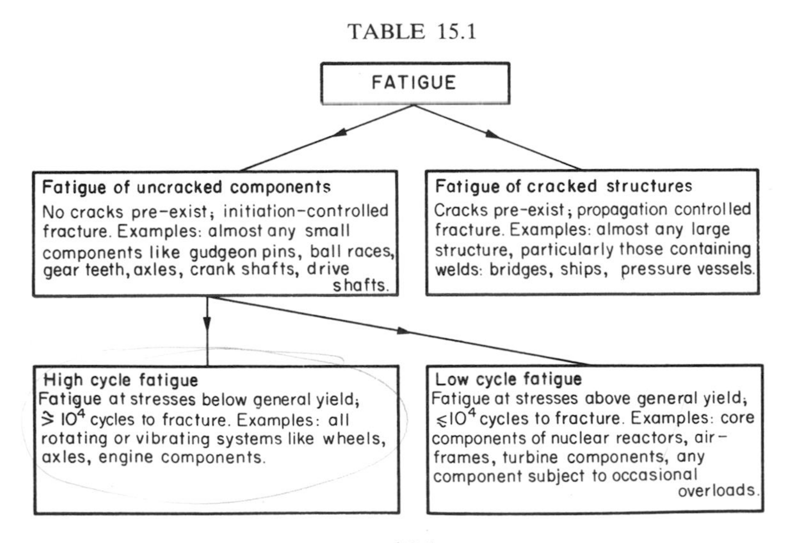

Fatigue behaviour of uncracked components

Tests are carried out by cycling the material either in tension (compression) or in rotating bending (Fig. 15.1). The stress, in general, varies sinusoidally with time, though modern servo-hydraulic testing machines allow complete control of the wave shape.

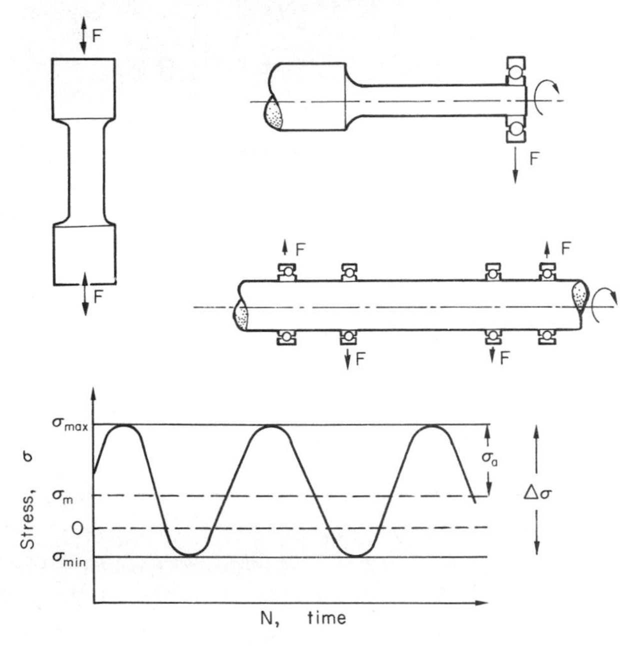

Fig. 15.1. Fatigue testing.

We define:

$$\Delta\sigma = \sigma_{max} - \sigma_{min}; \qquad \sigma_m = \frac{\sigma_{max} + \sigma_{min}}{2}; \qquad \sigma_a = \frac{\sigma_{max} - \sigma_{min}}{2}$$

where N = number of fatigue cycles and N_f = number of cycles to failure. We will consider fatigue under zero mean stress ($\sigma_m = 0$) first, and later generalise the results to non-zero mean stress.

For *high-cycle fatigue of uncracked components,* where neither σ_{max} nor $|\sigma_{min}|$ are above the yield stress, it is found empirically that the experimental data can be fitted to an equation of form

$$\Delta\sigma N_f^a = C_1. \tag{15.1}$$

This relationship is called *Basquin's Law.* Here, a is a constant (between $\frac{1}{8}$ and $\frac{1}{15}$ for most materials) and C_1 is a constant also.

For *low-cycle fatigue of un-cracked components* where σ_{max} or $|\sigma_{min}|$ are above σ_y, Basquin's Law no longer holds, as Fig. 15.2 shows. But a linear plot is obtained if the plastic strain range $\Delta\varepsilon^{pl}$, defined in Fig. 15.3, is plotted, on logarithmic scales, against

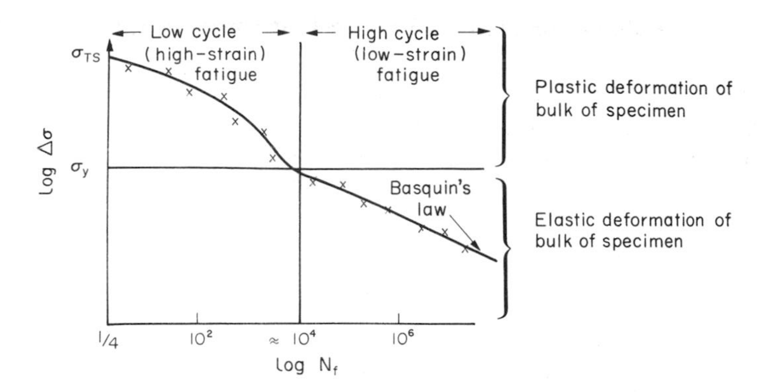

Fig. 15.2. Initiation-controlled high-cycle fatigue—Basquin's Law.

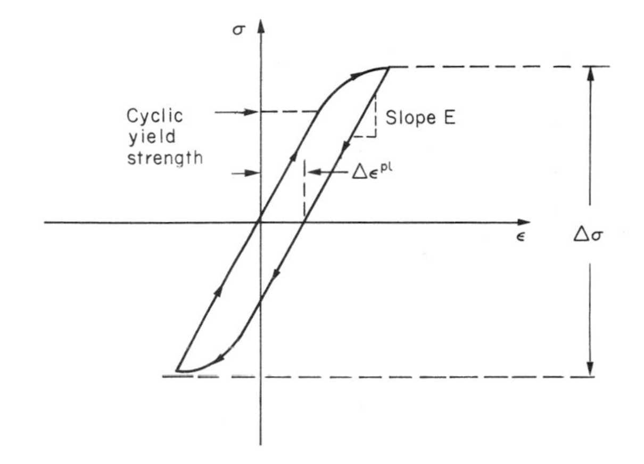

Fig. 15.3. The plastic strain range, $\Delta\varepsilon^{pl}$, in low-cycle fatigue.

the cycles to failure, N_f (Fig. 15.4). This result is known as the *Coffin–Manson Law*:

$$\Delta\varepsilon^{pl} N_f^b = C_2 \tag{15.2}$$

where b (0.5 to 0.6) and C_2 are constants.

These two laws (given data for a, b, C_1 and C_2) adequately describe the fatigue failure of unnotched components, cycled at constant amplitude about a mean stress of zero. What do we do when $\Delta\sigma$, and σ_m, vary?

When material is subjected to a mean tensile or compressive stress (i.e. σ_m is no longer zero) the stress range must be decreased to preserve the same N_f according to *Goodman's Rule* (Fig. 15.5)

$$\Delta\sigma_{\sigma m} = \Delta\sigma_0 \left(1 - \frac{|\sigma_m|}{\sigma_{TS}}\right). \tag{15.3}$$

(Here $\Delta\sigma_0$ is the cyclic stress range for failure in N_f cycles under zero mean stress, and $\Delta\sigma_{\sigma m}$ is the same thing for a mean stress of σ_m.) Goodman's Rule is empirical, and does

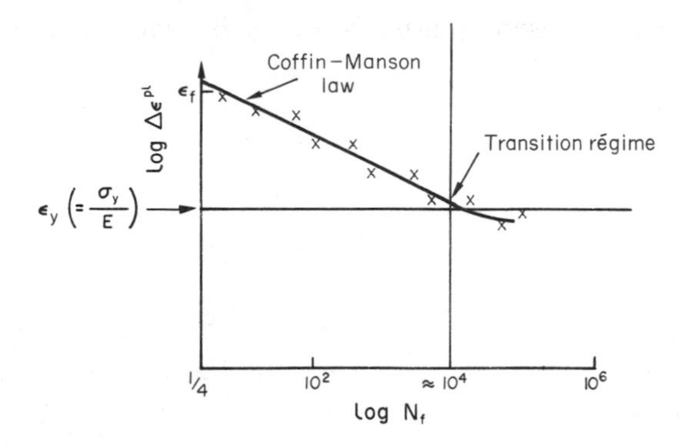

Fig. 15.4. Initiation-controlled low-cycle fatigue—the Coffin–Manson Law.

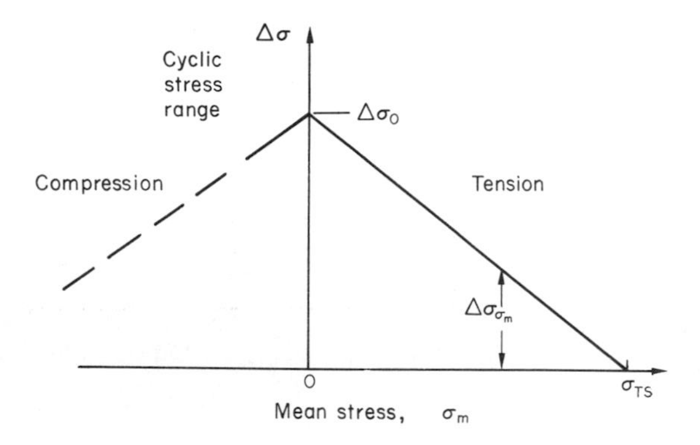

Fig. 15.5. Goodman's Rule—the effect of a non-zero mean stress on initiation-controlled fatigue.

not always work—then tests simulating service conditions must be carried out, and the results used for the final design. But preliminary designs are usually based on this rule.

When, in addition, $\Delta\sigma$ varies during the lifetime of a component, the approach adopted is to sum the damage according to *Miner's Rule* of *cumulative damage*:

$$\sum_i \frac{N_i}{N_{fi}} = 1. \tag{15.4}$$

Here N_{fi} is the number of cycles to fracture under the stress cycle in region i, and N_i/N_{fi}

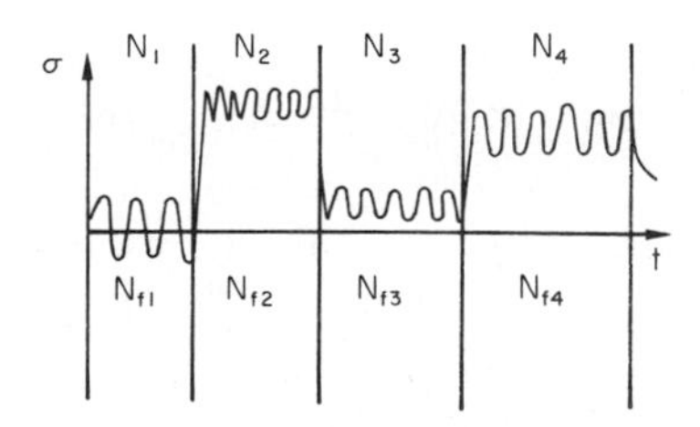

Fig. 15.6. Summing damage due to initiation-controlled fatigue.

is the fraction of the lifetime used up after N_i cycles in that region. Failure occurs when the sum of the fractions is unity (eqn. (15.4)). This rule, too, is an empirical one. It is widely used in design against fatigue failure; but if the component is a critical one, Miner's Rule should be checked by tests simulating service conditions.

Fatigue behaviour of cracked components

Large structures—particularly welded structures like bridges, ships, oil rigs, nuclear pressure vessels—always contain cracks. All we can be sure of is that the initial length of these cracks is less than a given length—the length we can reasonably detect when we check or examine the structure. To assess the safe life of the structure we need to know how long (for how many cycles) the structure can last before one of these cracks grows to a length at which it propagates catastrophically.

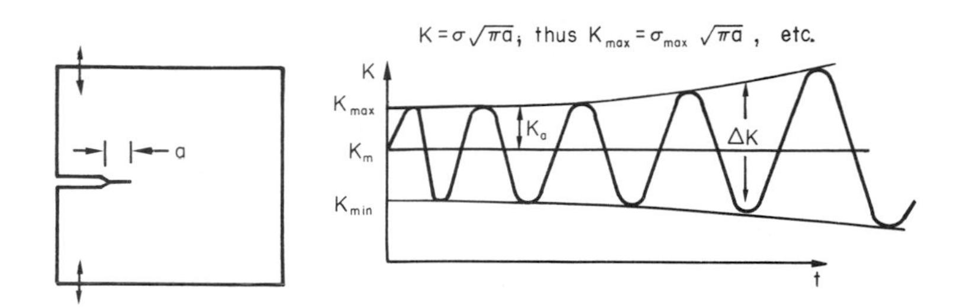

Fig. 15.7. Fatigue-crack growth in pre-cracked components.

Data on fatigue crack propagation are gathered by cyclically loading specimens containing a sharp crack like that shown in Fig. 15.7. We define

$$\Delta K = K_{max} - K_{min}; \qquad K_a = \frac{K_{max} - K_{min}}{2}; \qquad K_m = \frac{K_{max} + K_{min}}{2}.$$

The cyclic stress intensity ΔK increases with time (at constant load) because the crack grows. It is found that the crack growth per cycle, $\mathrm{d}a/\mathrm{d}N$, increases with ΔK in the way shown in Fig. 15.8.

In the steady-state régime, the crack growth rate is described by

$$\frac{\mathrm{d}a}{\mathrm{d}N} = A\Delta K^m \tag{15.5}$$

where A and m are material constants. Obviously, if a_0 (the initial crack length) is given, and the final crack length (a_f) at which the crack becomes unstable and runs rapidly is known or can be calculated, then the safe number of cycles can be estimated by integrating the equation

$$N_f = \int_0^{N_f} \mathrm{d}N = \int_{a_0}^{a_f} \frac{\mathrm{d}a}{A(\Delta K)^m}, \tag{15.6}$$

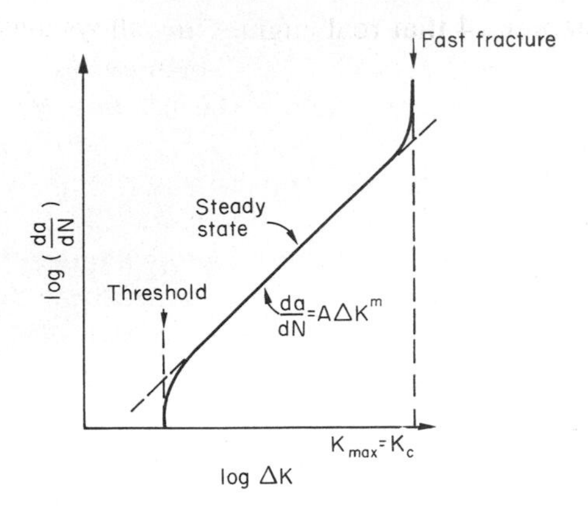

Fig. 15.8. Fatigue crack-growth rates for pre-cracked material.

remembering that $\Delta K = \Delta\sigma\sqrt{\pi a}$. Case Study 3 of Chapter 16 gives a worked example of this method of estimating fatigue life.

Fatigue mechanisms

Cracks grow in the way shown in Fig. 15.9. In a pure metal or polymer (left-hand diagram), the tensile cycle produces a plastic zone (Chapter 14) which makes the crack tip stretch open by the amount δ, creating new surface there. The compressive cycle squeezes the crack shut, and the new surface folds forward, extending the crack (roughly, by δ). On the next tensile cycle the same thing happens again, and the crack inches forward, roughly at $da/dN \approx \delta$.

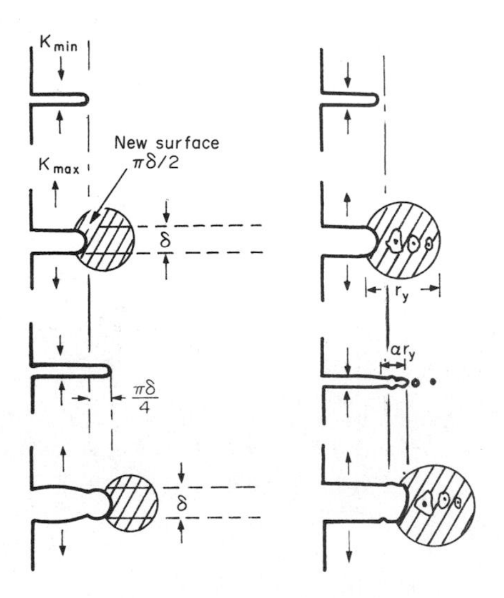

Fig. 15.9. How fatigue cracks grow.

We mentioned in Chapter 14 that real engineering alloys always have little inclusions in them. Then (right-hand diagram of Fig. 15.9), within the plastic zone, holes form and link with each other, and with the crack tip. The crack now advances a little faster than before, aided by the holes.

In *pre-cracked structures* these processes determine the fatigue life. In uncracked components subject to *low-cycle fatigue*, the general plasticity quickly roughens the surface, and a crack forms there, propagating first along a slip plane ("Stage 1" crack) and then, by the mechanism we have described, normal to the tensile axis (Fig. 15.10).

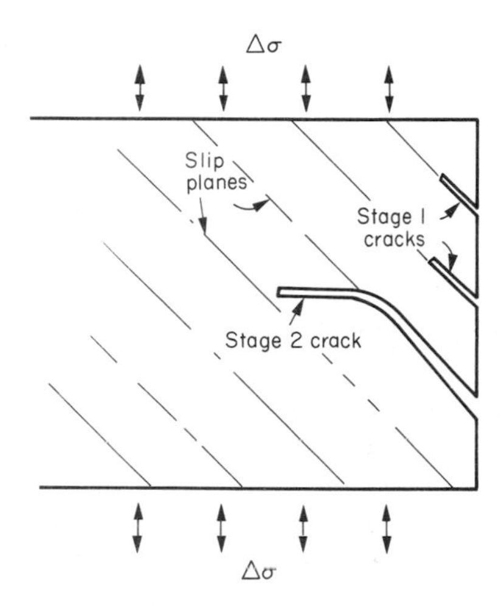

Fig. 15.10. How cracks form in low-cycle fatigue. Once formed, they grow as shown in Fig. 15.9.

High-cycle fatigue is different. When the stress is below general yield, almost all of the life is taken up in initiating a crack. Although there is no *general* plasticity, there is *local* plasticity wherever a notch or scratch or change of section concentrates stress. A crack ultimately initiates in the zone of one of these stress concentrations (Fig. 15.11) and

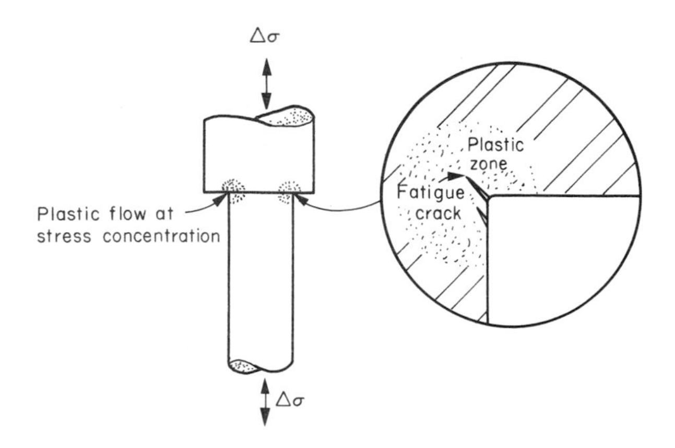

Fig. 15.11. How cracks form in high-cycle fatigue.

propagates, slowly at first, and then faster, until the component fails. For this reason, sudden changes of section or scratches are very dangerous in high-cycle fatigue, often reducing the fatigue life by orders of magnitude.

Further reading

R. W. Hertzberg, *Deformation and Fracture Mechanics of Engineering Materials*, Wiley, 1976, Chaps. 12 and 13.
J. F. Knott, *Fundamentals of Fracture Mechanics*, Butterworths, 1973, Chap. 9.
T. V. Duggan and J. Byrne, *Fatigue as a Design Criterion*, Macmillan, 1977.

CHAPTER 16

CASE STUDIES IN FAST FRACTURE AND FATIGUE FAILURE

Introduction

In this third set of Case Studies we examine three instances in which failure by crack-propagation was, or could have become, a problem. The first is the analysis of a die that failed by fast fracture the first time it was used. The second concerns a common problem: the checking, for safety reasons, of cylinders designed to hold gas at high pressure. The last is a fatigue problem: the safe life of a reciprocating engine known to contain a large crack.

Case study 1: fast fracture of a die

The die shown in Fig. 16.1 is designed for making superconducting alloys. The alloys are made by mixing the ingredients, in the form of metal powders, and compressing them in the die under as high a pressure as possible. This gives a slug of metal which is then sintered (heated at high temperature to make the particles adhere to one another)

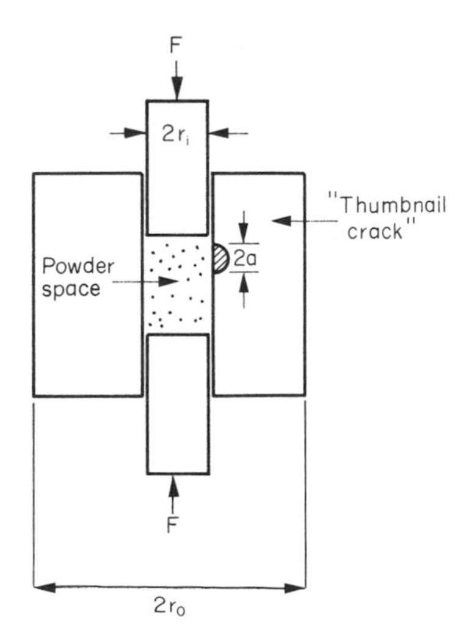

Fig. 16.1. Schematic view of powder-pressing die in operation.

Fig. 16.2. A powder-pressing die which failed suddenly by fast fracture.

and drawn down to fine wire. The higher the die pressure, the denser is the slug, and the better the final product. In an attempt to raise the pressure a new die was made from a special steel which could be heat-treated (heated and then quenched in oil) to give a very high yield strength. The intention was to use this high yield strength to permit very high working pressures.

It failed, on the first run, at about half the "safe" design load (Fig. 16.2). Examination of the fracture surface showed a thumbnail crack starting at the inner surface; it is visible in Fig. 16.3. You are asked to determine the cause of failure and recommend a solution.

Dimensions of the die, and material properties

The dimensions of the die are shown on Fig. 16.1. The important ones are

$$r_i = 6.4\ \text{mm},$$

$$r_0 = 38\ \text{mm},$$

$$a = 1.2\ \text{mm}.$$

A series of hardness tests showed that the average hardness of the material of the die was

$$H = 612\ \text{kg mm}^{-2}$$

$$= (9.81 \times 612) = 6000\ \text{MN m}^{-2}.$$

Fig. 16.3. A close-up of the fracture surface of Fig. 16.2, showing the "thumbnail" crack from which fast fracture took place.

Since $\sigma_y = H/3$ (Chapter 8), we have that

$$\sigma_y = 2000 \text{ MN m}^{-2}.$$

A sample of the same steel ("medium carbon chromium steel"), heat-treated to this hardness, had a toughness of

$$K_c = 22 \text{ MN m}^{-3/2}.$$

We already begin to see what has happened. If we look at Table 8.1 of yield strengths (Chapter 8), we see that σ_y for this steel is immensely high; but from what we said in Chapter 14, a high yield strength leads to a very low toughness, because crack blunting is suppressed by the difficulty of making plastic flow occur at the crack tip (Fig. 16.4).

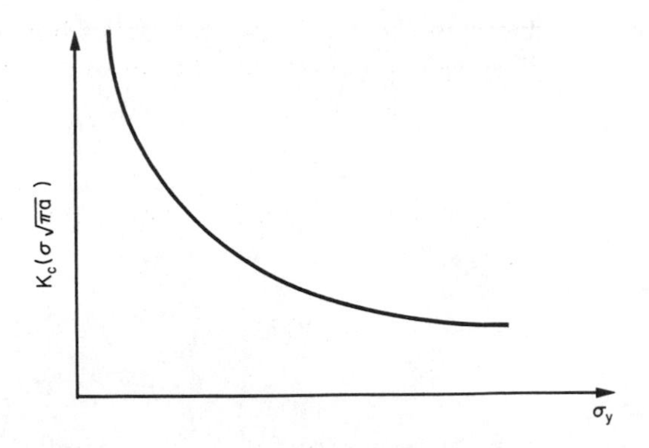

Fig. 16.4. The harder a material, the lower is its fracture toughness.

And indeed K_c for this steel *is* low: Table 13.1 in Chapter 13 shows that most steels have values of K_c between 50 and 200 MN m$^{-3/2}$.

Mechanics of die loading

The stresses in a thick-walled cylindrical pressure vessel (Fig. 16.5) are given in Mechanics texts (see **Further reading**) as

$$\left.\begin{aligned}
\sigma_t &= p\frac{\left(\dfrac{1}{r^2}+\dfrac{1}{r_0^2}\right)}{\left(\dfrac{1}{r_i^2}-\dfrac{1}{r_0^2}\right)} \quad \text{(a tensile stress)},\\
\sigma_r &= -p\frac{\left(\dfrac{1}{r^2}-\dfrac{1}{r_0^2}\right)}{\left(\dfrac{1}{r_i^2}-\dfrac{1}{r_0^2}\right)} \quad \text{(a compressive stress)}.
\end{aligned}\right\} \qquad (16.1)$$

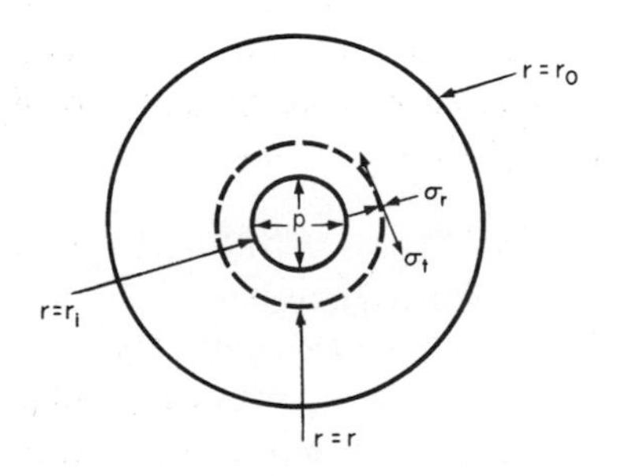

Fig. 16.5. Stresses in a thick-walled pressure vessel.

(Prove to yourself that σ_t reduces to the familiar value pr/t when $t=(r_0-r_i)$ is small compared with r_i.) From the first of these equations, the tensile stress, tending to open the crack, evaluated at $r=r_i$, gives

$$\sigma_t = 1.06p.$$

Designing against *plastic yield*, and with a safety factor of 3, the die could be used up to a pressure of

$$p = \frac{\sigma_y}{1.06 \times 3} \approx 630 \text{ MN m}^{-2}$$

and this was authorised.

It failed at about one-half of this pressure, by fast fracture from the thumbnail crack. If we calculate the die pressure at which a flaw of size $a = 1.2$ mm just becomes unstable we can understand why:

$$\sigma\sqrt{\pi a} = 1.06p\sqrt{\pi a} = K_c,$$

$$p = \frac{K_c}{1.06\sqrt{\pi a}} = 338 \text{ MN m}^{-2}.$$

Conclusions and recommendations

The steel was improperly heat-treated, leading to very low K_c; and the steel contained abnormally large defects (probably due to "hydrogen cracking"). The failure load for fast fracture was much less (by a factor of 6) than that for general yield; yet the design had been based on yield criteria alone.

The answer is *either* to heat-treat the steel (following suppliers' specifications) to raise K_c, at some loss of σ_y; *or* (better) to make the die from a pressure-vessel steel like HY100 for which $K_c = 150$ MN m$^{-3/2}$ and $\sigma_y = 1500$ MN m^{-2}.

CASE STUDY 2: COMPRESSED AIR TANKS FOR A SUPERSONIC WIND TUNNEL

The supersonic wind tunnels in the Aerodynamic Laboratory at Cambridge University are powered by a bank of twenty large cylindrical pressure vessels. Each time the tunnels are used, the vessels are slowly charged by compressors, and then quickly discharged through a tunnel. How should we go about designing and checking pressure vessels of this type to make sure they are safe?

Criteria for design of safe pressure vessels

First, the pressure vessel must be safe from plastic collapse: that is, the stresses must everywhere be below general yield. Second, it must not fail by fast fracture: if the largest cracks it could contain have length $2a$ (Fig. 16.6), then the stress intensity

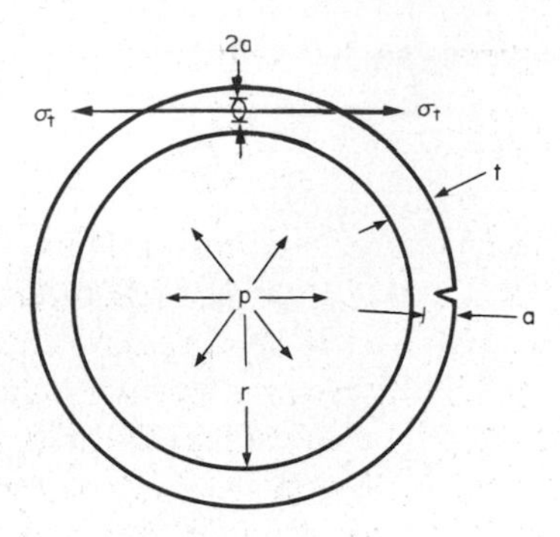

Fig. 16.6. Cracks in the wall of a pressure vessel.

$K \approx \sigma\sqrt{\pi a}$ must everywhere be less than K_c Finally, it must not fail by fatigue: the slow growth of a crack to the critical size at which it runs.

The stress σ_t in the wall of a cylindrical pressure vessel containing gas at pressure ρ is given by eqn. (16.1); or, if the wall is thin ($t \ll r$), by

$$\sigma = \frac{pr}{t}.$$

For general yielding,

$$\sigma = \sigma_y.$$

For fast fracture,

$$\sigma\sqrt{\pi a} = K_c.$$

Failure by general yield or fast fracture

Figure 16.7 shows the loci of general yielding and fast fracture plotted against crack size. The yield locus is obviously independent of crack size, and is simply given by

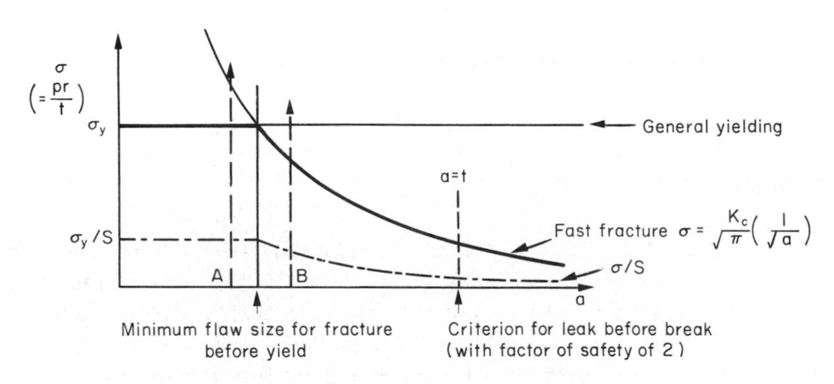

Fig. 16.7. Fracture modes for a cylindrical pressure vessel.

$\sigma = \sigma_y$. The locus of fast fracture can be written as

$$\sigma = \frac{K_c}{\sqrt{\pi}}\left(\frac{1}{\sqrt{a}}\right),$$

which gives a curved relationship between σ and a. If we pressurise our vessel at point A on the graph, the material will yield *before* fast fracture; this yielding can be detected by strain gauges and disaster avoided. If we pressurise at point B, fast fracture will occur at a stress less than σ_y, without warning and with catastrophic consequences; the point where the two curves cross defines a critical flaw size at which fracture by general yield and by fast fracture coincide. Obviously, if we know that the size of the largest flaw in our vessel is less than this critical value, our vessel will be safe (although we should also, of course, build in an appropriate safety factor S as well—as shown by the dash—dot line on Fig. 16.7).

Figure 16.8 shows the general yield and fast fracture loci for a pressure-vessel steel and an aluminium alloy. The critical flaw size in the steel is $\approx$9 mm; that in the aluminium alloy is $\approx$1 mm. It is easy to detect flaws of size 9 mm by ultrasonic testing, and pressure-vessel steels can thus be accurately tested non-destructively for safety—vessels with cracks larger than 9 mm would not be passed for service. Flaws of 1 mm size cannot be measured so easily or accurately, and thus aluminium is less safe to use.

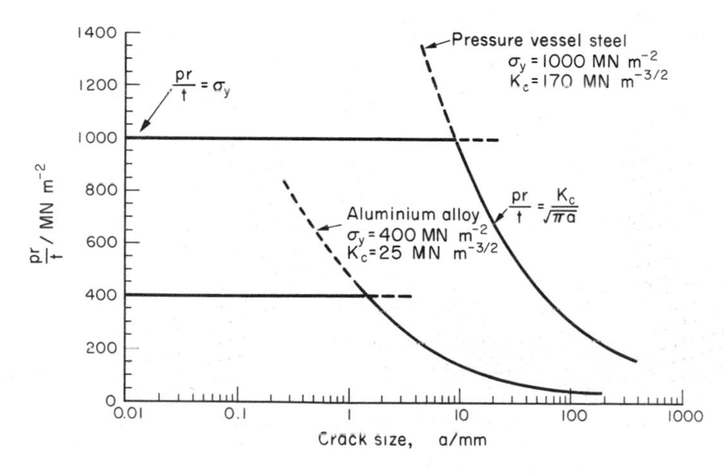

Fig. 16.8. Design against yield and fast fracture for a cylindrical pressure vessel.

Failure by fatigue

In the case of a pressure vessel subjected to *cyclic loading* (as here) cracks can grow by fatigue and a vessel initially passed as safe may subsequently become unsafe due to this crack growth. The probable extent of crack growth can be determined by making fatigue tests on pre-cracked pieces of steel of the same type as that used in the pressure vessel, and the safe vessel lifetime can be estimated by the method illustrated in Case Study 3.

Extra safety: leak before break

It is worrying that a vessel which is safe when it enters service may become unsafe by slow crack growth—either by fatigue or by stress corrosion (Chapter 23). If the consequences of catastrophic failure are very serious, then additional safety can be gained by designing the vessel so that it will *leak before it breaks* (like the partly inflated balloon of Chapter 13). Leaks are easy to detect, and a leaking vessel can be taken out of service and repaired. How do we formulate this leak-before-break condition?

If the critical flaw size for fast fracture is less than the wall thickness (t) of the vessel, then fast fracture can occur with no warning. But suppose the critical size ($2a_{crit}$) is greater than t—then gas will leak out through the crack before the crack is big enough to run. To be on the safe side we shall take

$$2a_{crit} = 2t.$$

The stress is defined by

$$\sigma\sqrt{\pi a_{crit}} = K_c$$

so that the permissible stress is

$$\sigma = \frac{K_c}{\sqrt{\pi t}}$$

as illustrated on Fig. 16.7.

There is, of course, a penalty to be paid for this extra safety: either the pressure must be lowered, or the section of the pressure vessel increased—often substantially.

Fig. 16.9. A pressure vessel in action—the boiler of the articulated steam locomotive *Merddin Emrys*, built in 1879 and still hauling passengers on the Festiniog narrow-gauge railway in North Wales.

Pressure testing

In many applications a pressure vessel may be tested for safety simply by hydraulic testing to a pressure that is higher—typically 1.5 to 2 times higher—than the normal operating pressure. Steam boilers (Fig. 16.9) are tested in this way, usually once a year. If failure does not occur at twice the working pressure, then the normal operating stress is at most one-half that required to produce fast fracture. If failure *does* occur under hydraulic test nobody will get hurt because the stored energy in compressed water is small. Periodic testing is vital because cracks in a steam boiler will grow by fatigue, corrosion, stress corrosion and so on. The procedure is safe only because cracks in applications of this type tend to grow slowly.

Case study 3: the safety of the Stretham engine

The Stretham steam pumping engine (Fig. 16.10) was built in 1831 as part of an extensive project to drain the Fens for agricultural use. In its day it was one of the largest beam engines in the Fens, having a maximum power of 105 horsepower at 15 rpm (it could lift 30 tons of water per revolution, or 450 tons per minute); it is now the sole surviving steam pump of its type in East Anglia.*

The engine could still be run for demonstration purposes. Suppose that you are called in to assess its safety. We will suppose that a crack 2 cm deep has been found in the connecting rod—a cast-iron rod, 21 feet long, with a section of 0.04 m^2. Will the crack grow under the cyclic loads to which the connecting rod is subjected? And what is the likely life of the structure?

Mechanics

The stress in the crank shaft is calculated approximately from the power and speed as follows. Bear in mind that approximate calculations of this sort may be in error by up to a factor of 2—but this makes no difference to the conclusions reached below. Referring to Fig. 16.11:

$$\text{Power} = 105 \text{ horsepower}$$
$$= 7.8 \times 10^4 \text{ J s}^{-1},$$
$$\text{Speed} = 15 \text{ rpm} = 0.25 \text{ rev s}^{-1},$$
$$\text{Stroke} = 8 \text{ feet} = 2.44 \text{ m},$$

$$\text{Force} \times 2 \times \text{stroke} \times \text{speed} \approx \text{power},$$

$$\therefore \text{Force} \approx \frac{7.8 \times 10^4}{2 \times 2.44 \times 0.25} \approx 6.4 \times 10^4 \text{ N}.$$

Nominal stress in the connecting rod $= F/A = 6.4 \times 10^4/0.04 = 1.6 \text{ MN m}^{-2}$ approximately.

* Until a couple of centuries ago much of the eastern part of England which is now called East Anglia was a vast area of desolate marshes, or fens, which stretched from the North Sea as far inland as Cambridge.

Fig. 16.10. Part of the Stretham steam pumping engine. In the foreground are the crank and the lower end of the connecting rod. Also visible are the flywheel (with separate spokes and rim segments, all pegged together), the eccentric drive to the valve-gear and, in the background, an early treadle-driven lathe for on-the-spot repairs.

Failure by fast fracture

For cast iron, $K_c = 18\ \mathrm{MN\ m^{-3/2}}$.

First, could the rod fail by fast fracture? The stress intensity is:

$$K = \sigma\sqrt{\pi a} = 1.6\sqrt{\pi.0.02}\ \mathrm{MN\ m^{-3/2}} = 0.40\ \mathrm{MN\ m^{-3/2}}.$$

It is so much less than K_c that there is no risk of fast fracture, even at peak load.

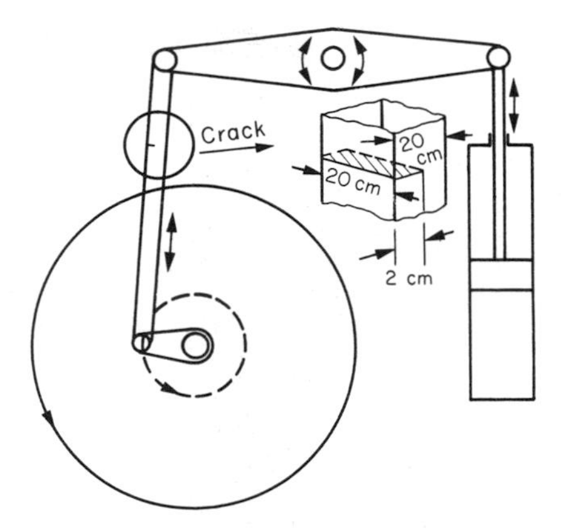

Fig. 16.11. Schematic of the Stretham engine.

Failure by fatigue

The growth of a fatigue crack is described by

$$\frac{da}{dN} = A(\Delta K)^m. \tag{16.2}$$

For cast iron,

$$A = 4.3 \times 10^{-8}\ \text{m}\ (\text{MN m}^{-3/2})^{-4},$$
$$m = 4.$$

We have that

$$\Delta K = \Delta\sigma\sqrt{\pi a}$$

where $\Delta\sigma$ is the range of the stress (Fig. 16.12). Although $\Delta\sigma$ is constant (at constant

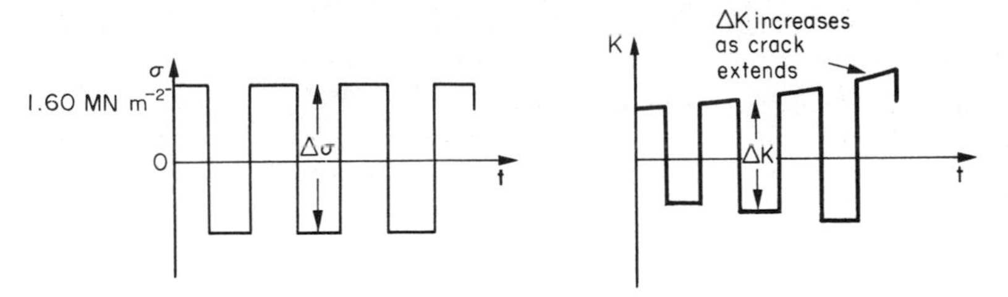

Fig. 16.12. Crack growth by fatigue in the Stretham engine.

power and speed), ΔK increases as the crack grows. Substituting in eqn. (16.2) gives

$$\frac{da}{dN} = A\Delta\sigma^4\pi^2a^2$$

and

$$dN = \frac{1}{(A\Delta\sigma^4\pi^2)}\frac{da}{a^2}.$$

Integration gives the number of cycles to grow the crack from a_1 to a_2:

$$N = \frac{1}{(A\Delta\sigma^4\pi^2)}\left\{\frac{1}{a_1} - \frac{1}{a_2}\right\}$$

for a range of a small enough that the crack geometry does not change appreciably.* Let us work out how long it would take our crack to grow from 2 cm to 3 cm. Then

$$N = \frac{1}{4.3\times10^{-8}(3.20)^4\pi^2}\left\{\frac{1}{0.02} - \frac{1}{0.03}\right\}$$
$$= 3.7\times10^5 \text{ cycles.}$$

This is sufficient for the engine to run for 8 hours on each of 52 open days for demonstration purposes, i.e. to give 8 hours of demonstration each weekend for a year. A crack of 3 cm length is still far too small to go critical, and thus the engine will be perfectly safe after a further 3.7×10^5 cycles. Under demonstration the power delivered will be far less than the full 105 horsepower, and because of the $\Delta\sigma^4$ dependence of N, the number of cycles required to make the crack grow to 3 cm might be as much as 30 times the one we have calculated.

The estimation of the total lifetime of the structure is more complex—substantial crack growth will make the crack geometry change significantly, and this will have to be allowed for in the calculations.

Conclusion and recommendation

A simple analysis shows that the engine is likely to be safe for limited demonstration use for a considerable period. After this period, continued use can only be sanctioned by regular inspection of the growing crack, or by using a more sophisticated analysis.

Further reading

J. F. Knott, *Fundamentals of Fracture Mechanics*, Butterworths, 1973.
T. V. Duggan and J. Byrne, *Fatigue as a Design Criterion*, Macmillan, 1977.
R. W. Hertzberg, *Deformation and Fracture Mechanics of Engineering Materials*, Wiley, 1976.
S. P. Timoshenko and J. N. Goodier, *Theory of Elasticity*, 3rd edition, McGraw Hill, 1970.

*See note at end of Chapter 13.

E. Creep deformation and fracture

CHAPTER 17

CREEP AND CREEP FRACTURE

Introduction

So far we have concentrated on mechanical properties at room temperature. Many structures—particularly those associated with energy conversion, like turbines, reactors, steam and chemical plant—operate at much higher temperatures.

As the temperature is raised, materials under loads which caused no permanent deformation at room temperature start to creep. Creep is slow, continuous deformation with time: the strain, instead of depending only on the stress, now depends on temperature and time also:

$$\varepsilon = f(\sigma, t, T) \quad \textbf{creeping solid.}$$

This is in contrast to the room-temperature behaviour of most metals and ceramics, in which the strain is, for practical purposes, independent of time:

$$\varepsilon = f(\sigma) \quad \textbf{elastic/plastic solid.}$$

It is common to refer to the former behaviour as "high-temperature" behaviour, and the latter as "low-temperature".

But what is a "high" temperature and what is a "low" temperature? Tungsten, used for lamp filaments, has a very high melting point—well over 3000°C. Room temperature, for tungsten, is a very low temperature. If made hot enough, however, tungsten will creep—that is the reason that lamps ultimately burn out. Tungsten lamps run at about 2000°C—this, for tungsten, is a high temperature. If you examine a lamp filament which has failed, you will see that it has sagged under its own weight until the turns of the coil have touched—that is, it has deformed by creep.

Figure 17.1 and Table 17.1 give melting points for metals and ceramics and softening temperatures for polymers. Most metals and ceramics have high melting points and, because of this, they start to creep only at temperatures well above room temperature—this is why creep is a less familiar phenomenon than elastic or plastic deformation. But lead, for instance, has a melting point of 600 K; at room temperature, it is at exactly half its absolute melting point. Room temperature for lead is a high temperature, and it creeps—as Fig. 17.2 shows. And the ceramic, ice, melts at 0°C. "Temperate" glaciers (those close to 0°C) are at a temperature at which ice creeps rapidly—that is why glaciers move. Even the thickness of the Antarctic ice cap, which controls the levels of the earth's oceans, is determined by the creep-spreading of the ice at about −30°C.

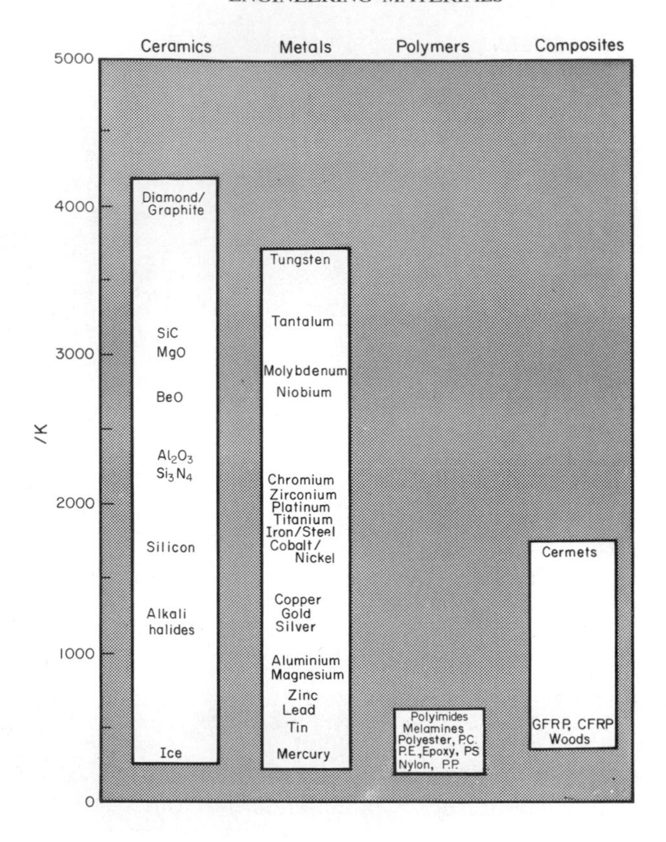

Fig. 17.1. Melting or softening temperature.

The point, then, is that the temperature at which materials start to creep depends on their melting point. As a general rule, it is found that creep starts when

$$T > 0.3 \text{ to } 0.4\ T_M \text{ for metals,}$$

$$T > 0.4 \text{ to } 0.5\ T_M \text{ for ceramics,}$$

where T_M is the melting temperature in degrees kelvin. However, special alloying procedures can raise the temperature at which creep becomes a problem.

Polymers, too, creep—many of them do so at room temperature. As we said in Chapter 5, most common polymers are not crystalline, and have no well-defined melting point. For them, the important temperature is the glass temperature, T_G, at which the Van der Waals bonds solidify. Well above this temperature, the polymer is in a leathery or rubbery state, and will creep under load. Well below, it becomes hard (and sometimes brittle) and, for practical purposes, no longer creeps.

We shall examine first the form that creep data take, and then rank materials roughly in order of their resistance to creep.

TABLE 17.1
MELTING OR SOFTENING[(S)] TEMPERATURE

Material	/K	Material	/K
Diamond, graphite	4000	Silica glass	1100
Tungsten	3680	Aluminium	933
Tantalum	3250	Magnesium	923
Silicon carbide, SiC	3110	Soda glass	700–900
Magnesia, MgO	3073	Zinc	692
Molybdenum	2880	Polyimides	580–630[(S)]
Niobium	2740	Lead	600
Beryllia, BeO	2700	Tin	505
Alumina, Al_2O_3	2323	Melamines	400–480[(S)]
Silicon nitride, Si_3N_4	2173	Polyesters	450–480[(S)]
Chromium	2148	Polycarbonates	400[(S)]
Zirconium	2125	Polyethylene, high-density	300[(S)]
Platinum	2042	Polyethylene, low-density	360[(S)]
Titanium	1943	Foamed plastics, rigid	300–380[(S)]
Iron	1809	Epoxy, general purpose	340–380[(S)]
Cobalt	1768	Polystyrenes	370–380[(S)]
Nickel	1726	Nylons	340–380[(S)]
Cermets	1700	Polyurethane	365[(S)]
Silicon	1683	Acrylic	350[(S)]
Alkali halides	800–1600	GFRP	340[(S)]
Uranium	1405	CFRP	340[(S)]
Copper	1356	Polypropylene	330[(S)]
Gold	1336	Ice	273
Silver	1234	Mercury	235

Fig. 17.2. Lead pipes often creep noticeably over the years.

Creep testing and creep curves

Creep tests require careful temperature control. Typically, a specimen is loaded in tension or compression, usually at constant load, inside a furnace which is maintained at a constant temperature, T. The extension is measured as a function of time. Figure 17.3 shows a typical set of results from such a test. Metals, polymers and ceramics all show creep curves of this general shape.

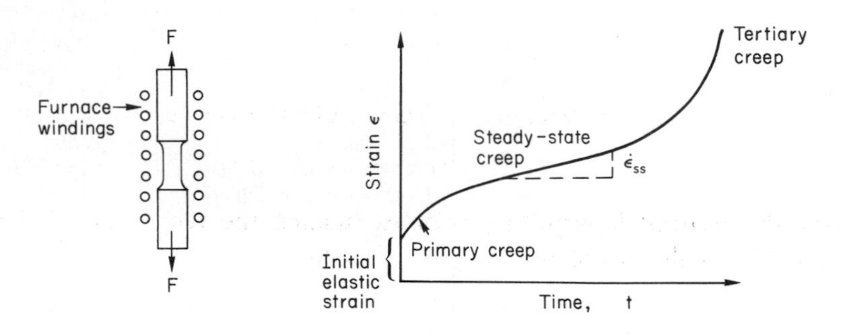

Fig. 17.3. Creep testing and creep curves.

Although the initial and the primary creep strain cannot be neglected, they occur quickly, and they can be treated in much the way that elastic deflection is allowed for in a structure. But thereafter, the material enters steady-state, or secondary creep, and the strain increases steadily with time. In designing against creep, it is usually this steady accumulation of strain with time that concerns us most.

By plotting the log of the steady creep-rate, $\dot{\varepsilon}_{ss}$, against log(stress), at constant T, as shown in Fig. 17.4 we can establish that

$$\dot{\varepsilon}_{ss} = B\sigma^n \qquad (17.1)$$

where n, the *creep exponent*, usually lies between 3 and 8. This sort of creep is called "*power-law*" *creep*. (At low σ, a different régime is entered where $n \approx 1$; we shall discuss this low-stress deviation from power-law creep in Chapter 19, but for the moment we shall not comment further on it.)

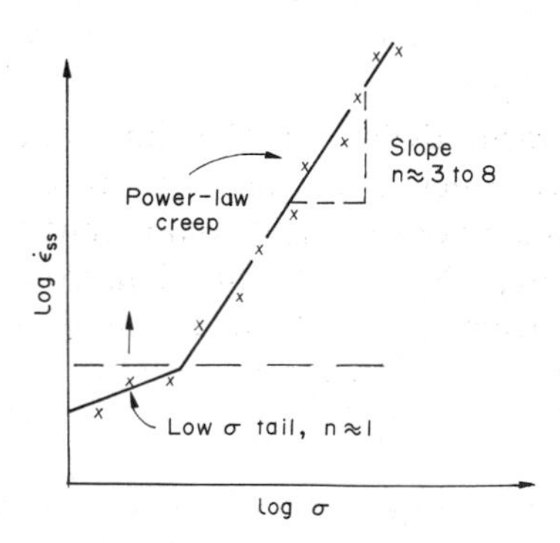

Fig. 17.4. Variation of creep rate with stress.

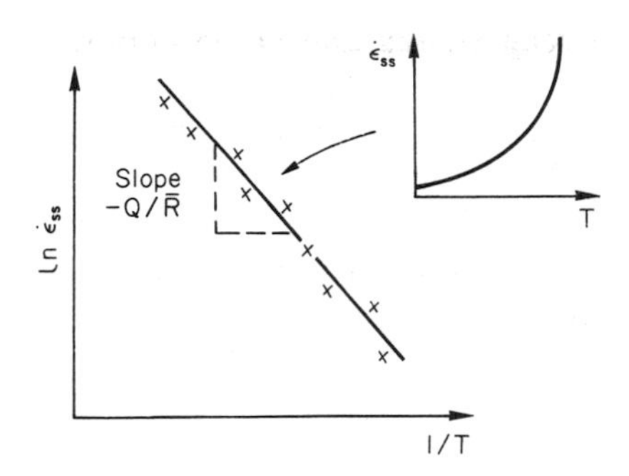

Fig. 17.5. Variation of creep rate with temperature.

By plotting the *natural* logarithm (ln) of $\dot{\varepsilon}_{ss}$ against the reciprocal of the *absolute* temperature ($1/T$) as shown in Fig. 17.5, we find that:

$$\dot{\varepsilon}_{ss} = Ce^{-(Q/\bar{R}T)}. \tag{17.2}$$

Here $\bar{R}$ is the Universal Gas Constant ($8.31\ \mathrm{J\,mol^{-1}\,K^{-1}}$) and Q is called the *Activation Energy for Creep*—it has units of $\mathrm{J\,mol^{-1}}$. Note that the creep rate increases exponentially with temperature (Fig. 17.5, inset). An increase in temperature of 20°C can *double* the creep rate.

Combining these two dependences of $\dot{\varepsilon}_{ss}$ gives, finally,

$$\dot{\varepsilon}_{ss} = A\sigma^n e^{-(Q/\bar{R}T)} \tag{17.3}$$

where A is the creep constant. The values of the constants A, n and Q naturally vary from material to material, and have to be found experimentally.

Consequences of power-law creep

(a) *At constant stress (or load), strain accumulates with time.* Ice in the Antarctic ice cap and in glaciers spreads and flows by power-law creep. Polymers distort at and above room temperature. Metals and ceramics in high-temperature equipment and structures (reactors; turbines) suffer slow strain.

(b) *At constant displacement, stress relaxes with time.* Bolts in turbine casings, etc., must be regularly tightened.

The relaxation time (arbitrarily defined as the time taken for the stress to relax to half its original value) can be calculated from the power-law creep data as follows. Consider a bolt which is tightened onto a rigid component so that the initial stress in its shank is σ_i. In this geometry (Fig. 17.6) the length of the shank must remain constant—that is, the *total* strain in the shank ε^{tot} must remain constant. But creep strain ε^{cr} can *replace* elastic strain ε^{el}, causing the stress to relax. At any time t

$$\varepsilon^{tot} = \varepsilon^{el} + \varepsilon^{cr}. \tag{17.4}$$

But

$$\varepsilon^{el} = \sigma/E.$$

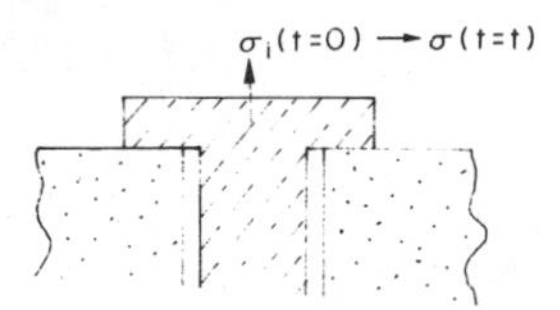

Fig. 17.6. Schematic of a strained bolt at high temperature.

And (at constant temperature)

$$\dot{\varepsilon}^{cr} = B\sigma^n.$$

Since ε^{tot} is constant, we can differentiate eqn. (17.4) with respect to time and substitute the other two equations into it to give

$$\frac{1}{E}\frac{d\sigma}{dt} = -B\sigma^n. \qquad (17.5)$$

Integrating from $\sigma = \sigma_i$ at $t = 0$ to $\sigma = \sigma$ at $t = t$ gives

$$\frac{1}{\sigma^{n-1}} - \frac{1}{\sigma_i^{n-1}} = (n-1)BEt. \qquad (17.6)$$

Figure 17.7 shows how the initial elastic strain σ_i/E is slowly replaced by creep strain, and the stress in the bolt relaxes. If, as an example, it is a casing bolt in a large turbo generator, it will have to be retightened at intervals to prevent steam leaking from the turbine. The time interval between retightening, t_r, can be calculated by evaluating the time it takes for σ to fall to (say) one-half of its initial value. Setting $\sigma = \sigma_i/2$ and rearranging gives

$$t_r = \frac{(2^{n-1} - 1)}{(n-1)BE\sigma_i^{n-1}}. \qquad (17.7)$$

Experimental values for n, A and Q for the material of the bolt thus enable us to decide how often the bolt will need retightening. Note that overtightening the bolt does not help because t_r falls rapidly as σ_i increases.

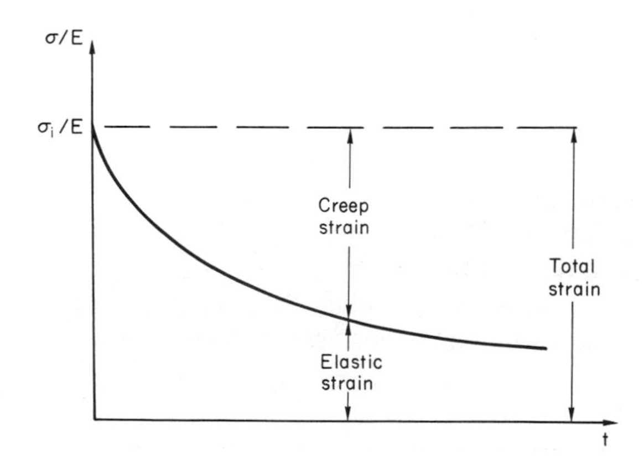

Fig. 17.7. Replacement of elastic strain by creep strain with time at high temperature.

Creep damage and creep fracture

During creep, damage, in the form of internal cavities, accumulates. The damage first appears at the start of the Tertiary Stage of the creep curve and grows at an increasing rate thereafter. The shape of the Tertiary Stage of the creep curve reflects this: as the holes grow, the section of the sample decreases, and (at constant load) the stress goes up. Since $\dot{\varepsilon} \propto \sigma^n$, the creep rate goes up even faster than the stress does (Fig. 17.8).

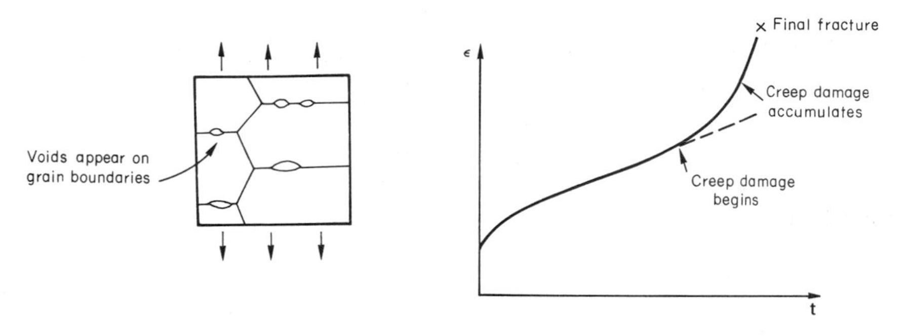

Fig. 17.8. Creep damage.

In many high-strength alloys this creep damage appears early in life and leads to failure after small creep strains (as little as 1%). In high-temperature design it is important to make sure:

(a) That the *creep strain* ε^{cr} during the design life is acceptable.

(b) That the *creep ductility* ε_f^{cr} (strain to failure) is adequate to cope with the acceptable creep strain.

(c) That the *time to failure*, t_f, at the design loads and temperatures is longer (by a suitable safety factor) than the design life.

Times to failure are normally presented as *creep-rupture* diagrams (Fig. 17.9). Their application is obvious.

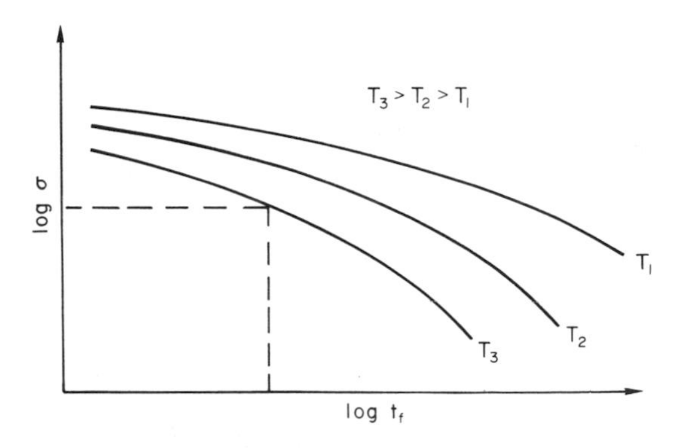

Fig. 17.9. Creep-rupture diagram.

Creep-resistant materials

From what we have said so far it should be obvious that the first requirement that we should look for in choosing materials that are resistant to creep is that they should have high melting (or softening) temperatures. If the material can then be used at less than 0.3 of its melting temperature creep will not be a problem. If it has to be used above this temperature, various alloying procedures can be used to increase creep resistance. To understand these, we need to know more about the mechanisms of creep—the subject of the next two chapters.

Further reading

I. Finnie and W. R. Heller, *Creep of Engineering Materials*, McGraw Hill, 1959.
J. Hult, *Creep in Engineering Structures*, Blaisdell, 1966.

CHAPTER 18

KINETIC THEORY OF DIFFUSION

Introduction

We saw in the last chapter that the rate of steady-state creep, $\dot{\varepsilon}_{ss}$, varies with temperature as

$$\dot{\varepsilon}_{ss} = Ce^{-(Q/\bar{R}T)};$$

here Q is the activation energy for creep (J mol^{-1} or, more usually, kJ mol^{-1}), $\bar{R}$ is the universal gas constant (8.31 J mol^{-1} K^{-1}) and T is the absolute temperature (K). This is an example of *Arrhenius's Law*—a rate-law which has great generality. It applies not only to the rate of creep, but to the rate of oxidation (Chapter 21), of corrosion (Chapter 23), of diffusion (this chapter), even to the rate at which bacteria multiply and milk turns sour. It states that the *rate increases exponentially with temperature* (or that the time for a given amount of creep, or of oxidation, *decreases* exponentially with temperature) as Fig. 18.1 shows. And if the rate of any process which follows

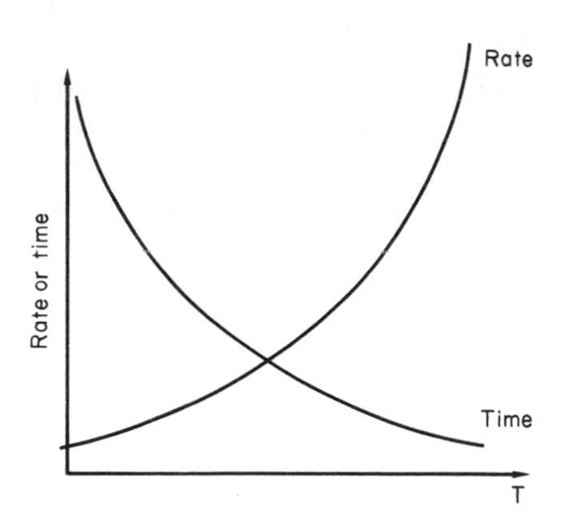

Fig. 18.1. Consequences of Arrhenius's Law.

Arrhenius's Law is plotted on a ln scale against $1/T$, a straight line with a slope of $-Q/\bar{R}$ is obtained (Fig. 18.2).

In this chapter we discuss the origin of Arrhenius's Law and its application to *diffusion*. In the next, we examine how it is that the rate of diffusion determines that of creep.

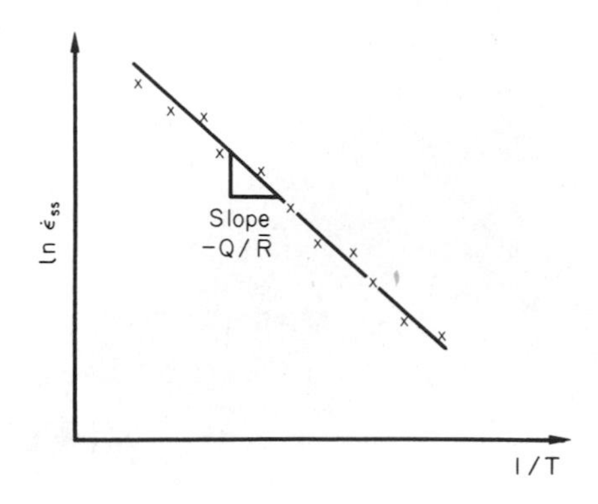

Fig. 18.2. Creep rates follow Arrhenius's Law.

Diffusion and Fick's Law

First, what do we mean by *diffusion*? If we take a dish of water and add a drop of ink to it, the ink will spread sideways into the water. Provided the water is stagnant, the spreading of the ink is due to the movement of "ink" molecules by random exchanges with the water molecules. It occurs in such a sense that the "ink" molecules move from regions where they are concentrated to regions where they are less concentrated. In other words, the ink diffuses down the ink *concentration gradient.* This behaviour is described by *Fick's first law of diffusion.*

$$J = -D\frac{dc}{dx}. \tag{18.1}$$

J is the number of ink molecules diffusing down the concentration gradient *per second per unit area*; it is termed the *flux* of molecules (Fig. 18.3). c is the concentration of ink molecules in the water, defined as the *number* of ink molecules per *unit volume* of the ink–water solution. D is the *diffusion coefficient* for ink water—it has units of $m^2 s^{-1}$. This diffusive behaviour, naturally, is not just limited to ink in water—it occurs in all other liquids, and more remarkably, in all solids as well. For example, the alloy brass consists of a mixture of zinc in copper; diffusion of zinc atoms can occur through the solid copper in just the way that ink can diffuse through water. Because we are concerned here with creep in solids, we shall now confine ourselves to talking about diffusion in the solid state also.

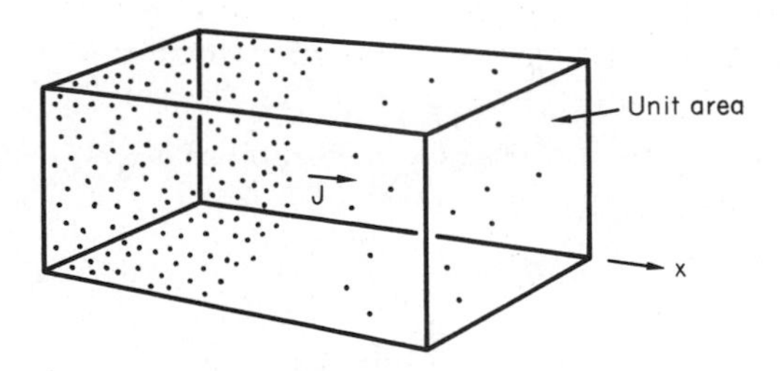

Fig. 18.3. Diffusion down a concentration gradient.

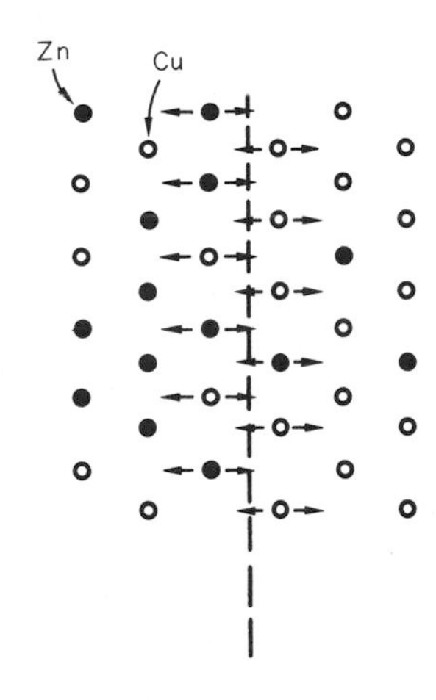

Fig. 18.4. Atom jumps across a plane.

Physically, diffusion occurs because atoms, even in a solid, are able to move—to jump from one atomic site to another. Figure 18.4 shows a solid in which there is a concentration gradient of black atoms: there are more to the left of the broken line than there are to the right. If atoms jump across the broken line at random, then there will be a *net flux* of black atoms to the right (simply because there are more on the left to jump), and, of course, a net flux of white atoms to the left. We shall now analyse these fluxes in more detail to derive Fick's Law.

The atoms in a solid vibrate, or oscillate, about their mean positions, with a frequency ν (typically about $10^{13}\ s^{-1}$). The crystal lattice defines these mean positions. At a temperature T, the average energy (kinetic plus potential) of an atom is $3kT$ where k is Boltzmann's constant (1.38×10^{-23} J atom^{-1} K^{-1}). But this is only the average energy. As atoms (or molecules) vibrate, they collide, and energy is continually transferred from one to another. Although the *mean* energy is $3kT$, at any instant, there is a certain probability that an atom has more or less than this. It can be shown from statistical mechanical theory that the probability, p, that an atom will have, at any instant, an energy $\geqslant q$ is

$$p = e^{-q/kT}.$$

Let us now see how this expression is relevant to the diffusion of zinc in copper for example. We shall take two adjacent lattice planes in our brass, having two slightly different zinc concentrations as shown in exaggerated form in Fig. 18.5. Let us denote these two planes as A and B. Now for the zinc to *diffuse* from A to B, down our *concentration gradient*, zinc atoms have to "squeeze" between the copper atoms (this is a very simplified statement—we shall elaborate on it later on in this chapter). An alternative way of expressing this is that they have to overcome an *energy* obstacle of height q, as shown in Fig. 18.5. Now, the number of zinc atoms in layer A is n_A, so that the number of zinc atoms that have enough energy to climb over the energy barrier from A to B at any instant is

$$n_A p = n_A e^{-q/kT}. \tag{18.2}$$

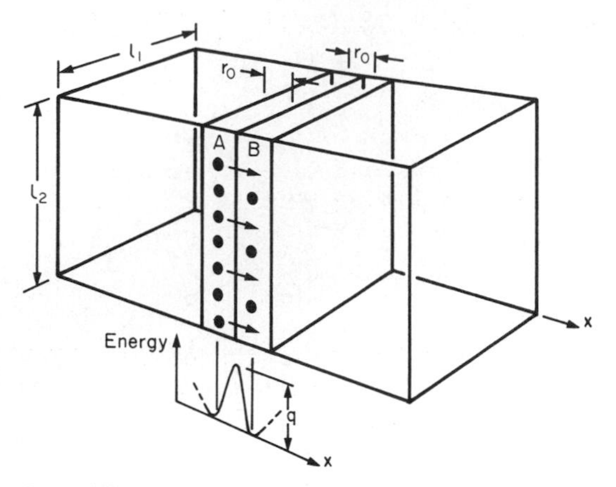

Fig. 18.5. Diffusion requires atoms to cross the energy barrier q.

In order for these atoms to *actually* climb over the barrier from A to B, they must of course be moving in the right direction. The number of times each zinc atom oscillates towards B is $\approx \nu/6$ per second (there are six possible directions in which the zinc atoms can move in three dimensions, only one of which is from A to B). Thus the *number* of atoms that *actually* jump from A to B per second is

$$\frac{\nu}{6} n_A e^{-q/kT}.$$

Similarly, if the number of zinc atoms in layer B is n_B, the number of zinc atoms that climb over the barrier from B to A per second is

$$\frac{\nu}{6} n_B e^{-q/kT}. \qquad (18.3)$$

Thus the net number of zinc atoms climbing over the barrier per second is

$$\frac{\nu}{6}(n_A - n_B) e^{-q/kT}, \qquad (18.4)$$

which represents our net movement of zinc atoms down the concentration gradient from A to B. The *flux* of atoms is

$$J = \nu \frac{(n_A - n_B)}{6 l_1 l_2} e^{-q/kT} \qquad (18.5)$$

from our previous definition.

From the definition of concentration we have

$$c_A = \frac{n_A}{l_1 l_2 r_0}, \qquad c_B = \frac{n_B}{l_1 l_2 r_0} \qquad (18.6)$$

where c_A and c_B are the zinc concentrations at A and B. Substituting for n_A and n_B in eqn. (18.5) gives

$$J = \frac{\nu}{6} r_0 (c_A - c_B) e^{-q/kT}. \qquad (18.7)$$

But $-(c_A - c_B)/r_o$ is simply dc/dx, so that eqn. (18.7) reduces to

$$J = -\frac{\nu r_0^2}{6} e^{-q/kT} \frac{dc}{dx}. \tag{18.8}$$

Comparison of this equation with eqn. (18.1) shows that it is simply Fick's First Law, with diffusion coefficient D determined as

$$D = \frac{\nu r_0^2}{6} e^{-q/kT}. \tag{18.9}$$

For most diffusing atoms q is an inconveniently small quantity and it is better to use the larger quantities $Q = N_A q$ and $\bar{R} = N_A k$ where N_A is Avogadro's number. In addition, $\nu r_0^2/6$ is usually written as D_0, giving, finally,

$$D = D_0 e^{-Q/\bar{R}T} \tag{18.10}$$

where D_0 is a constant having units of $m^2\,s^{-1}$.

This method of writing D emphasises the exponential dependence of D on temperature, and gives a conveniently sized activation energy (expressed as J per mole of diffusing atoms rather than per atom). The main thing about eqn. (18.10) is that the exponential dependence of D on temperature has exactly the same form as the dependence of $\dot{\varepsilon}_{ss}$ on temperature that we have been trying to explain.

Data for diffusion coefficients

Diffusion coefficients are usually measured by plating a thin layer of a *radioactive isotope* of the diffusing atoms or molecules onto the bulk material (for example, radioactive zinc onto copper). The temperature is raised to the diffusion temperature for a measured time, during which the isotope diffuses into the bulk. The sample is cooled and sectioned, and the concentration of isotope measured as a function of depth by measuring the radiation it emits. D_0 and Q are calculated from the diffusion profile. Materials Handbooks list data for D_0 and Q for various atoms diffusing in metals and ceramics (e.g. zinc in brass, carbon in steel, oxygen in MgO, etc.). Diffusion occurs in polymers and composites too, but there are still very few data.

TABLE 18.1
DATA FOR BULK SELF-DIFFUSION

Material class	$D_0/m^2\,s^{-1}$	$Q/\bar{R}T_M$
B.C.C. metals (W, Mo, Fe below 911°C, etc.)	1.6×10^{-4}	17.8
H.C.P. metals (Zn, Mg, Ti, etc.)	5×10^{-5}	17.3
F.C.C. metals (Cu, Al, Ni, Fe above 911°C, etc.)	5×10^{-5}	18.4
Alkali halides (NaCl, LiF, etc.)	2.5×10^{-3}	22.5
Oxides (MgO, FeO, Al_2O_3, etc.)	3.8×10^{-4}	23.4

It is found that, for a given *class of material* (e.g. f.c.c. metals, or refractory oxides) the diffusion parameter D_0 for mass transport—and this is the one that is important in creep—is almost constant; and that the activation energy is proportional to the melting temperature T_M (K) so that $Q/\bar{R}T_M$, too, is a constant (which is why creep is related to the melting point). This means that *many* diffusion problems can be solved approximately using the data given in Table 18.1.

Mechanisms of diffusion

In our discussion so far we have begged the question of just how the atoms in a solid move around when they diffuse. There are several ways in which this can happen. For simplicity, we shall talk only about crystalline solids, although of course diffusion occurs in amorphous solids as well.

Bulk diffusion: interstitial and vacancy diffusion

Diffusion in the bulk of the crystal can occur by two mechanisms. The first is *interstitial* diffusion. Atoms in all crystals have spaces, or *interstices*, between them, and *small* atoms dissolved in the crystal can diffuse around by squeezing between atoms, travelling from one interstice to another. Carbon, a small atom, diffuses through steel in this way; in fact C, O, N, B and H diffuse interstitially in most crystals (Fig. 18.6).

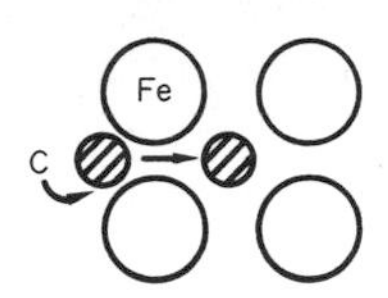

Fig. 18.6. Interstitial diffusion.

The second mechanism is that of *vacancy* diffusion. When zinc diffuses in brass, for example, the zinc atom (comparable in size to the copper atom) cannot fit into the interstices—the zinc atom has to wait until a *vacancy*, or missing atom, appears next to it before it can move. This is the mechanism by which most diffusion in crystals takes place (Figs. 18.7 and 10.4).

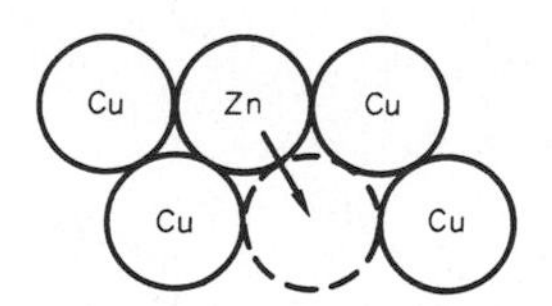

Fig. 18.7. Vacancy diffusion.

Fast diffusion paths: grain boundary and dislocation core diffusion

Diffusion in the bulk crystals may sometimes be *short circuited* by diffusion down grain boundaries or dislocation cores. The boundary acts as a planar channel, about 2 atoms wide, with a local diffusion rate which can be as much as 10^6 times greater than in the bulk (Figs. 18.8 and 10.4). The dislocation core, too, can act as a high

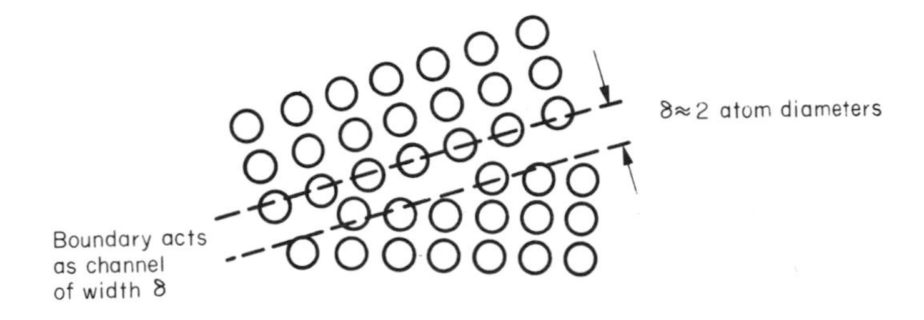

Fig. 18.8. Grain-boundary diffusion.

conductivity "wire" of cross-section about $(2b)^2$, where b is the atom size (Fig. 18.9). Of course, their contribution to the total diffusive flux depends also on how many grain boundaries or dislocations there are: when grains are small or dislocations numerous, their contribution is very important.

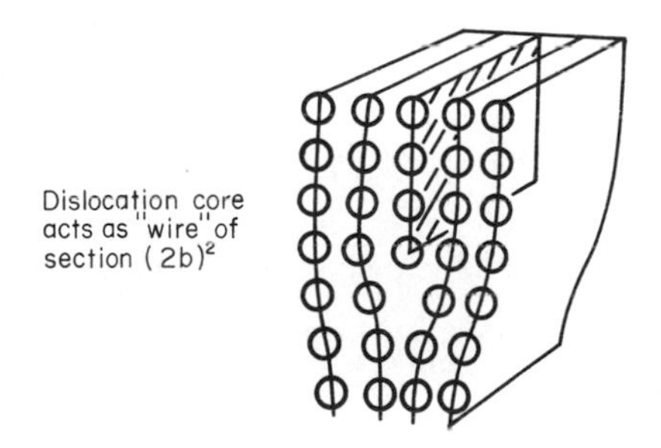

Fig. 18.9. Dislocation-core diffusion.

In Chapter 19 we shall see how the basic ideas of diffusion that we have been talking about here can be used to explain how creep takes place.

Further reading

P. G. Shewmon, *Diffusion in Solids*, McGraw Hill, 1963.
W. D. Kingery, *Introduction to Ceramics*, Wiley, 1960, Chap. 8.
G. H. Geiger and D. R. Poirier, *Transport Phenomena in Metallurgy*, Addison-Wesley, 1973, Chap. 13.
C. J. Smithells, *Metals Reference Book*, 5th edition, Butterworths, 1976 (for diffusion data).

CHAPTER 19

MECHANISMS OF CREEP, AND CREEP-RESISTANT MATERIALS

Introduction

In Chapter 17 we showed that, when a material is loaded at a high temperature, it creeps; i.e. it deforms, continuously and permanently, at a stress that is less than the stress that would cause any permanent deformation or yielding in a conventional tensile or compressive test. In order to understand how we can make engineering materials more resistant to creep deformation and creep fracture, we must first look at how creep takes place on an atomic level, i.e. we must identify and understand the *mechanisms* by which creep takes place.

There are two mechanisms of creep: *dislocation creep* (which gives power-law behaviour) and *diffusional creep* (which gives linear-viscous creep). The rate of both is usually limited by diffusion, so both follow Arrhenius's Law. Diffusion becomes appreciable at about $0.3T_M$—that is why materials start to creep above this temperature.

Creep mechanisms: metals and ceramics

Dislocation creep (giving power-law creep)

As we saw in Chapter 10, when a crystalline material is deformed plastically, the stress required to cause yielding is that needed to make the dislocations (a) overcome the intrinsic lattice resistance and (b) overcome the obstructing effect of obstacles (e.g. dissolved solute atoms, precipitates formed with undissolved solute atoms, or other dislocations). Diffusion of atoms can help to "unlock" dislocations from obstructions in their path, and the movement of these unlocked dislocations under the applied stress is what leads to dislocation creep.

How does this unpinning occur? Consider the case of a dislocation which cannot glide because a precipitate blocks its path (Fig. 19.1). The glide force (τb per unit length) is balanced by the reaction f_0 from the precipitate. But unless the dislocation hits the precipitate at its mid-plane (an unlikely event) there is a component of force left over. It is the component $\tau b \tan \theta$, which tries to push the dislocation *out of its slip plane.*

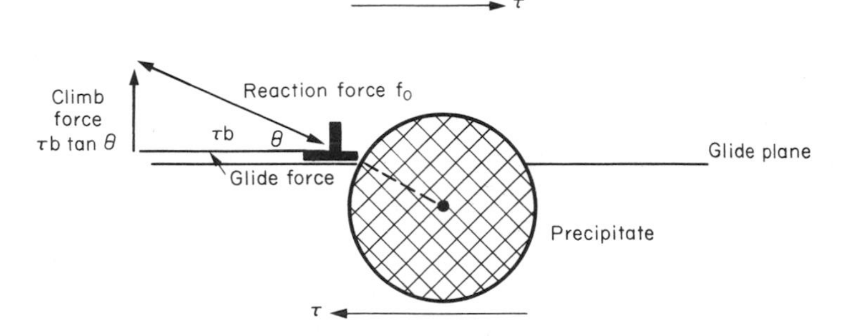

Fig. 19.1. The climb force on a dislocation.

The dislocation cannot *glide* upwards by the shearing of atom planes—the atomic geometry is wrong for the dislocation to work in this way—but the dislocation *can* move upwards if atoms at the bottom of the half-plane are able to diffuse away (Fig. 19.2). We have come across Fick's Law in which differences in *concentration* provided the driving force for diffusion. A *mechanical* force can do exactly the same thing, and this is what leads to the diffusion of atoms away from the "loaded" dislocation. The process is called "climb", and since it requires diffusion, it can occur only when the temperature is above $0.3T_M$ or so. At lower temperatures (0.3–$0.5T_M$) core diffusion tends to be the dominant mechanism; at higher temperatures it is bulk diffusion (Fig. 19.2).

This climb unlocks the dislocations from the precipitates which pin them and further slip can then take place (Fig. 19.3). Similar behaviour takes place for pinning by solute, and by other dislocations. Ultimately, of course, the unlocked dislocations glide up to adjacent obstacles, and the whole cycle of events takes place again. This explains the *progressive, continuous,* nature of creep; and the role of diffusion, with diffusion coefficient

$$D = D_0 e^{-Q/\bar{R}T}$$

explains the dependence of creep rate on *temperature*, with

$$\dot{\varepsilon}_{ss} = A\sigma^n e^{-Q/\bar{R}T}. \tag{19.1}$$

The dependence of creep rate on applied *stress* is obviously due to the climb force: the higher σ, the higher the climb force, the more dislocations become unlocked per second, the more dislocations glide per second, and the higher is the strain rate.

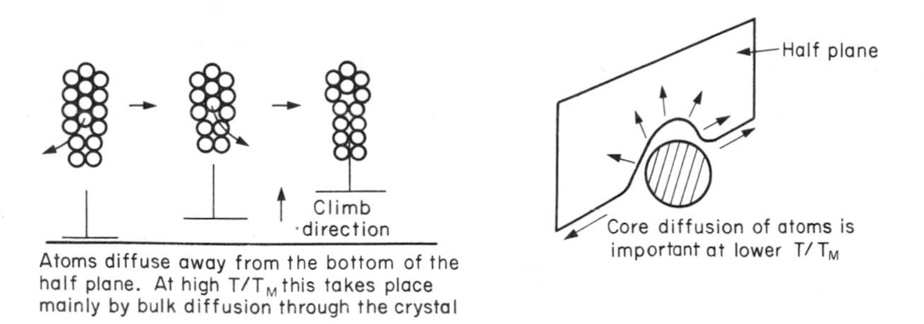

Fig. 19.2. How diffusion leads to climb.

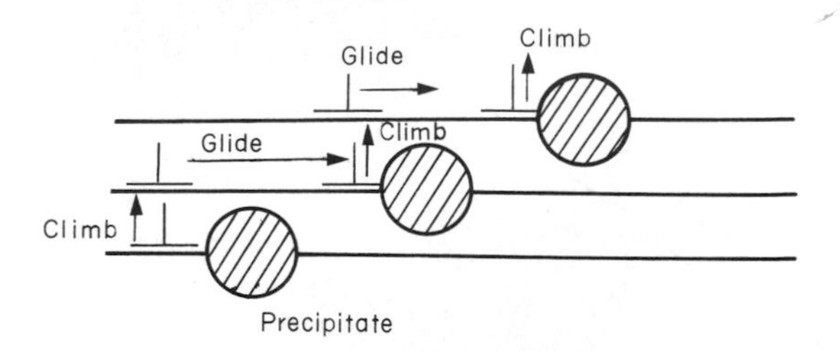

Fig. 19.3. How the climb–glide sequence leads to creep.

Diffusion creep (giving linear-viscous creep)

As the stress is reduced, the rate of power-law creep (eqn. (19.1)) falls quickly (remember n is between 3 and 8). But creep does not stop; instead, an alternative mechanism takes over. As Fig. 19.4 shows, the applied stress, σ, can be relieved by grain elongation; here, σ acts again as a mechanical driving force, this time for *diffusion* from one side of the grain to the other. At high T/T_M, this diffusion takes place through the crystal itself, by bulk diffusion. The rate of creep is then obviously proportional to the *diffusion coefficient D*, and to the stress σ (because σ drives diffusion in the same way that dc/dx does in Fick's Law); and the creep rate varies as $1/d^2$ where d is the grain size (because when d gets larger, matter has to diffuse further):

$$\dot{\varepsilon}_{ss} = C\frac{D\sigma}{d^2} = \frac{C'\sigma e^{-Q/RT}}{d^2} \tag{19.2}$$

where C and C' are constants. At lower T/T_M, when bulk diffusion is slow, grain-boundary diffusion takes over, but the creep rate is still proportional to σ. In order that holes do not open up between the grains, grain-boundary *sliding* is required as an accessory to this process.

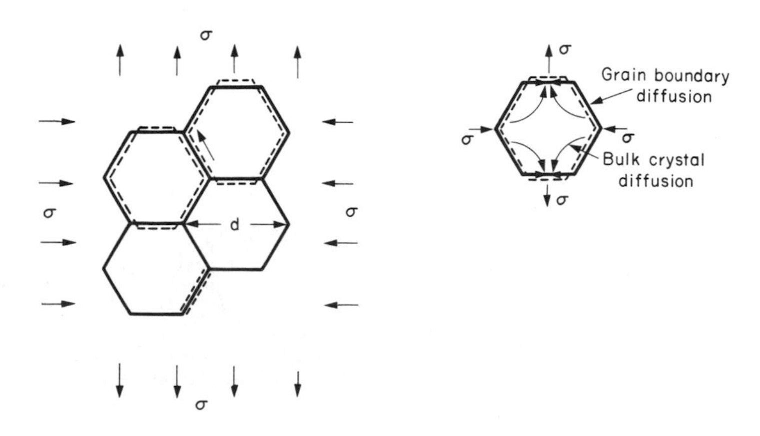

Fig. 19.4. How creep takes place by diffusion.

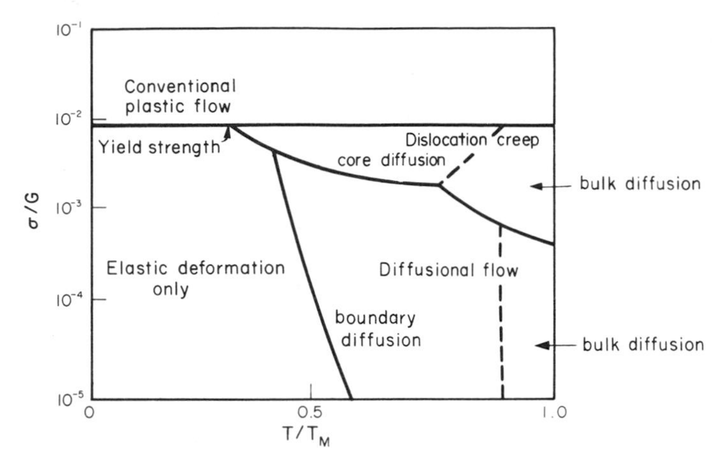

Fig. 19.5. Deformation mechanisms at different stresses and temperatures.

Deformation mechanism diagrams

This competition between mechanisms is conveniently summarised on Deformation Mechanism Diagrams (Figs. 19.5 and 19.6). They show the range of stress and temperature (Fig. 19.5) or of strain-rate and stress (Fig. 19.6) in which we expect to find each sort of creep (they also show where plastic yielding occurs, and where deformation is simply elastic). Diagrams like these are available for many metals and ceramics, and are a useful summary of creep behaviour, helpful in selecting a material for high-temperature applications.

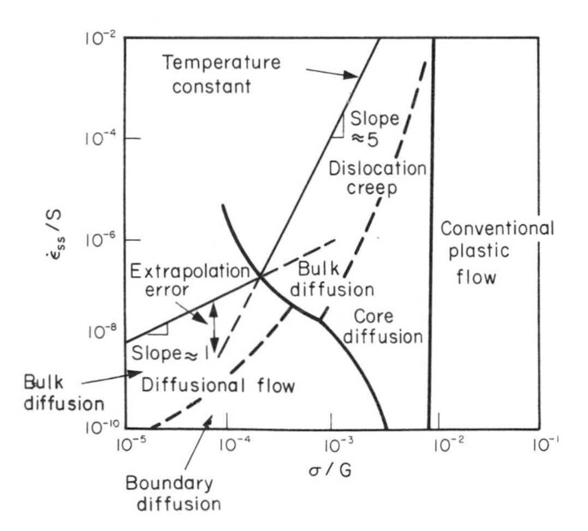

Fig. 19.6. Deformation mechanisms at different strain-rates and stresses.

Implications for design with metals and ceramics

How can we make use of this theoretical understanding of creep mechanisms to help us design creep-resistant materials?

Extrapolating test data

Many engineering components (e.g. tie bars in furnaces, super-heater tubes, high-temperature pressure vessels in chemical reaction plants) are expected to withstand moderate creep loads for long times (say 20 years) without significant straining. Obviously, one would like to be able to test new materials for these applications without having to run the tests for 20 years and more! It is thus tempting to speed up the tests by increasing the load to get observable creep in a short test time.

Now, if this procedure takes us across the boundary between two different types of mechanism, we shall have problems about extrapolating our test data to the operating conditions. Extrapolation based on power-law creep will be on the dangerous side as shown in Fig. 19.6.

Metal-forming operations

Sometimes creep *is* desirable. Extrusion, hot rolling, hot pressing and forging are carried out at temperatures at which power-law creep is the dominant mechanism of deformation. Then raising the temperature reduces the pressures or forces required for the operation. The change in forming pressure for a given change in temperature can be calculated from eqn. (19.1).

Designing metals and ceramics to resist power-law creep

If you are asked to select, or even to design, a material which will resist power-law creep, the criteria (all based on the ideas of this chapter and the last) are:

(a) Choose a material with a high melting point, since diffusion (and thus creep-rates) scale as T/T_M.
(b) Maximise obstructions to dislocation motion by alloying to give a solid solution and precipitates—as much of both as possible; the precipitates must, of course, be stable at the service temperature.
(c) Choose, if this is practical, a solid with a large lattice resistance: this means covalent bonding (as in many oxides, and in silicates, silicon carbide, silicon nitride, and related compounds).

Current creep-resistant materials are successful because they satisfy these criteria. Those you are most likely to come across are (in order of increasing T_M):

RR58: an aluminium alloy containing a solid solution and precipitates; low T_M; good only to 150°C, but low density.

High-alloy type 304, 316, 321 *stainless steels:* iron containing a solid solution (mainly Ni and Cr) and precipitates (carbides and intermetallics); good to 600°C.
Low-alloy ferritic steels: iron containing up to 4% of Cr, Mo and V, deriving most of their creep resistance from carbide precipitates; good to 650°C.
Nickel-based superalloys: a wide range of alloys of nickel containing a solid solution (mainly Cr, W, Co) and precipitates (carbides and intermetallics); good to 950°C (see Chapter 20).
Refractory oxides and carbides: notably alumina, Al_2O_3; glass ceramics based on SiO_2; silicon carbide, SiC; silicon nitride, Si_3N_4; and sialons—alloys of Si_3N_4 and Al_2O_3; potentially good to 1300°C (see Chapter 20), they make use of a high lattice resistance.

Designing metals and ceramics to resist diffusional flow

Diffusional flow is important when grains are small (as they often are in ceramics) and when the component is subject to high temperatures at low loads. To select a material which resists it, you should

(a) Choose a material with a high melting temperature.
(b) Arrange that it has a large grain size, so that diffusion distances are long and grain boundaries do not help diffusion much—single crystals are best of all.
(c) Arrange for precipitates at grain boundaries to impede grain-boundary sliding.

Metallic alloys are usually designed to resist power-law creep: diffusional flow is only rarely considered. One major exception is the range of directionally solidified ("DS") alloys described in the Case Study of Chapter 20: here special techniques are used to obtain very large grains.

Ceramics, on the other hand, often deform predominantly by diffusional flow (because their grains are small, and the high lattice resistance already suppresses power-law creep). Special heat treatments to increase the grain size can make them more creep-resistant.

Creep mechanisms: polymers

Creep of polymers is a major design problem. The glass temperature T_G of most polymers is close to room temperature. Well below T_G, the polymer is a glass (often containing crystalline regions—Chapter 5) and is a brittle, elastic solid—rubber, cooled in liquid nitrogen, is an example. Well above T_G the Van der Waals bonds within the polymer melt, and it becomes a rubber (if the polymer chains are cross-linked) or a viscous liquid (if they are not). Thermoplastics, which can be moulded when hot, are a simple example; well below T_G they are elastic; well above, they are Newtonian-viscous liquids.

Newtonian-viscous flow is a sort of creep. Like diffusion creep, its rate increases linearly with stress and exponentially with temperature, with

$$\dot{\varepsilon}_{ss} = C\sigma e^{-Q/\bar{R}T} \tag{19.3}$$

where Q is the activation energy for viscous flow.

The exponential term appears for the same reason as it does in diffusion; it describes the rate at which molecules can slide past each other, permitting flow. The molecules have a lumpy shape (see Fig. 5.9) and the lumps key the molecules together. The activation energy, Q, is the energy it takes to push one lump of a molecule past that of a neighbouring molecule. If we compare the last equation with that defining the *viscosity* (*for the tensile deformation of a viscous material*)

$$\eta = \frac{\sigma}{3\dot{\varepsilon}} \tag{19.4}$$

we see that the viscosity is

$$\eta = \frac{3}{C} e^{+Q/\bar{R}T}. \tag{19.5}$$

(The factor 3 appears because the viscosity is defined for shear deformation—as is the shear modulus G. For tensile deformation we want the viscous equivalent of Young's modulus E. The answer is 3η, for much the same reason that $E \approx (8/3)G$—see Chapter 3.)

Data giving C and Q for polymers are available from suppliers. Then eqn. (19.3) allows injection moulding or pressing temperatures and loads to be calculated.

But the temperature range in which most polymers are used is that near T_G when they are neither simple elastic solids nor viscous liquids; they are *visco-elastic* solids. If we represent the elastic behaviour by a spring and the viscous behaviour by a dash-pot, then visco-elasticity (at its simplest) is described by a coupled spring and dash-pot (Fig. 19.7). Applying a load causes creep, but at an ever-decreasing rate because the spring takes up the tension. Releasing the load allows slow reverse creep, caused by the extended spring.

Real polymers require more elaborate systems of springs and dash-pots to describe them. This approach of *polymer rheology* can be developed to provide criteria for design with structural polymers. At present, this is rarely done; instead, graphical data (showing the creep extension after time t at stress σ and temperature T) are used to provide an estimate of the likely deformation during the life of the structure.

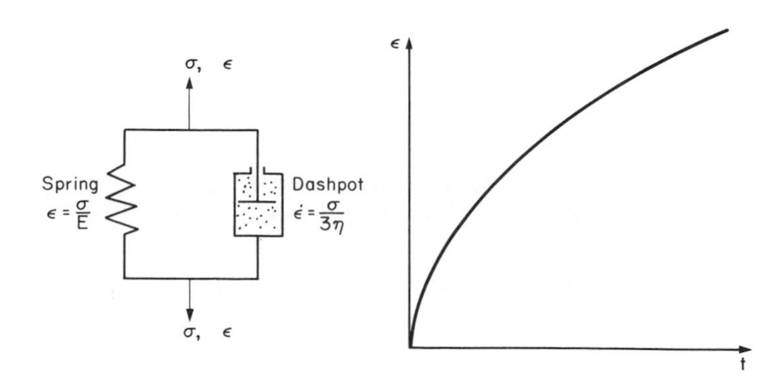

Fig. 19.7. A model to describe creep in polymers.

Implications for design with polymers

Selection of polymers

The glass temperature of a polymer increases with the degree of cross-linking; heavily cross-linked polymers (e.g. epoxies) are therefore more creep-resistant at room temperature than those which are less cross-linked (e.g. polyethylene). The viscosity of polymers above T_G increases with molecular weight, so that the rate of creep there is reduced by having a high molecular weight. Finally, crystalline or partly crystalline polymers (e.g. high-density polyethylene) are more creep-resistant than those which are entirely glassy (e.g. low-density polyethylene).

Designing polymers to resist creep

The creep-rate of polymers is reduced by filling them with glass or silica powders, roughly in proportion to the amount of filler added (PTFE on saucepans and polypropylene used for automobile components are both strengthened in this way). Much better creep resistance is obtained with composites containing continuous fibres (GFRP; and CFRP) because much of the load is now carried by the fibres which, being very strong, do not creep at all.

Further reading

I. Finnie and W. R. Heller, *Creep of Engineering Materials*, McGraw Hill, 1959.

CHAPTER 20

THE TURBINE BLADE—A CASE STUDY IN CREEP-LIMITED DESIGN

Introduction

In the last chapter we saw how a basic knowledge of the mechanisms of creep was an important aid to the development of materials with good creep properties. An outstanding example of the way in which this knowledge has been used is in the development of materials for the high-pressure stage of a modern aircraft gas turbine. In this Case Study we examine the properties such materials must have, the way in which the present generation of materials has evolved, and the likely direction of their future development.

As you may know, the *ideal* thermodynamic efficiency of a heat engine is given by

$$\frac{T_1 - T_2}{T_1} = 1 - \frac{T_2}{T_1} \tag{20.1}$$

where T_1 and T_2 are the absolute temperatures of the heat source and heat sink respectively. Obviously the greater T_1, the greater the maximum efficiency that can be derived from the engine. In practice the efficiency is a good deal less than this maximum value, but increases in combustion temperature in a turbofan engine will, nevertheless, generate corresponding increases in engine efficiency. Figure 20.1 shows the variation in efficiency of a turbofan engine plotted as a function of the turbine inlet temperature. In 1950 a typical engine operated at 700°C. There was, therefore, considerable incentive at that period to increase the inlet temperature because of the steepness of the fuel-consumption curve at that temperature. In 1975 the RB211 operated at 1350°C, with a 50% saving in fuel per unit power output over the 1950 engines. However, because of the shallowness of the consumption curve at 1400°C, further improvement of materials does not seem justified in terms of fuel savings. Why then the continued interest in materials development to enable turbines to function at still higher temperatures?

Well, there is a second factor—the *performance* of a given size of engine. Figure 20.2 shows a typical plot of the power output of a particular engine against turbine inlet temperature. This increases *linearly* with temperature, and obviously there is a continuing incentive to increase inlet temperatures in order to improve the power–weight ratio of aircraft, and reap the financial benefits of increased payload.

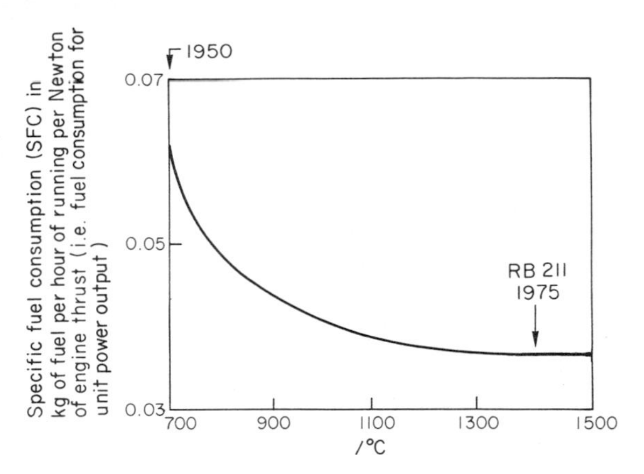

Fig. 20.1. Turbofan efficiency at different inlet temperatures.

Properties required of a turbine blade

Let us now examine the development of turbine-blade materials to meet the challenge of increasing engine temperatures. Although so far we have been stressing the need for excellent creep properties, a turbine-blade alloy must satisfy other criteria too. The criteria we need to satisfy altogether are listed in Table 20.1.

We shall talk about requirement (b) in Chapter 21. Toughness and fatigue we have discussed already—obviously a blade must be tough enough to withstand the impact of birds, pieces from a broken compressor blade, etc. In addition, differential expansion and contraction between different parts of the blade at different temperatures generate stresses which can crack a brittle material. Differential expansion and contraction can also lead to fatigue if they occur frequently, and the alloy must therefore be resistant to this thermally generated fatigue, or *thermal fatigue.* The alloy composition and structure must remain *stable* at high temperature—for example, precipitate particles can dissolve away if the alloy is overheated and the creep

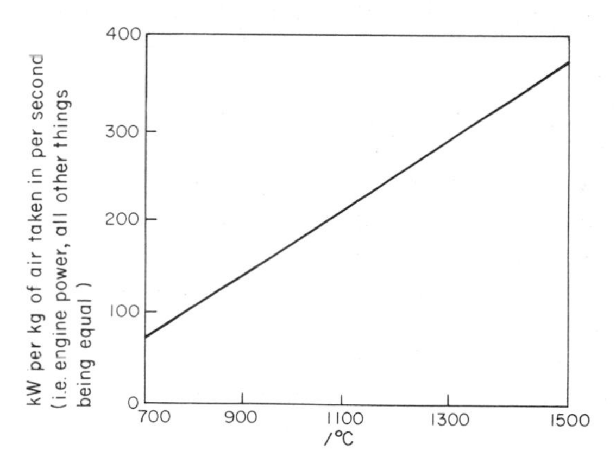

Fig. 20.2. Turbofan power at different inlet temperatures.

TABLE 20.1
ALLOY REQUIREMENTS

(a) Resistance to creep.
(b) Resistance to high-temperature oxidation.
(c) Toughness.
(d) Thermal fatigue resistance.
(e) Thermal stability.
(f) Low density.

properties will then fall off significantly. Finally, the *density* must be as low as possible—not so much because of blade weight but because of the need for stronger and hence heavier turbine discs to take the radial load.

These requirements severely limit our choice of creep-resistant materials. For example, ceramics, with their high softening temperatures and low densities, are ruled out for aero-engines because they are far too brittle (they are under evaluation for use in land-based turbines, where the risks and consequences of sudden failure are less severe—see below). Cermets offer no great advantage because their metallic matrices soften at much too low a temperature. The materials which at present best fill the bill are the nickel-based super-alloys.

Nickel-based super-alloys

A classic example of a material designed to be resistant to dislocation (power-law) creep at high stresses is the alloy used for turbine blades in the high-pressure stage of an aircraft turbofan engine. At take-off, the blade is subjected to a stress of $\approx 250\ \text{MN m}^{-2}$, and the design specification requires that this stress shall be supported for 30 hours at 850°C without more than a 0.1% irreversible creep strain. In order to meet these stringent requirements, an alloy based on nickel has evolved with the rather mind-boggling specification given in Table 20.2.

No one tries to remember exact details of this or similar alloys. But the point of all these complicated additions of foreign atoms to the nickel is straightforward. It is: (a) to have as many atoms in solid solution as possible (the cobalt; the tungsten; and the chromium); (b) to form stable, hard precipitates of compounds like Ni_3Al, Ni_3Ti, MoC,

TABLE 20.2
COMPOSITION OF TYPICAL CREEP-RESISTANT BLADE

	/wt. %		/wt. %
Ni	59	Mo	0.25
Co	10	C	0.15
W	10	Si	0.1
Cr	9	Mn	0.1
Al	5.5	Cu	0.05
Ta	2.5	Zr	0.05
Ti	1.5	B	0.015
Hf	1.5	S	<0.008
Fe	0.25	Pb	<0.0005

TaC to obstruct the dislocations; and (c) to form a protective surface oxide film of Cr_2O_3 (we shall discuss this in Chapter 22). Figure 20.3 (a and b) shows a piece of a nickel-based super-alloy cut open to reveal its complicated structure.

These super-alloys are remarkable materials. They resist creep so well that they can be used at 850°C—and since they melt at 1280°C, this is 0.72 of their (absolute) melting point. They are so hard that they cannot be machined easily by normal methods, and must be precision-cast to their final shape. This is done by *investment casting*: a precise wax model of the blade is embedded in an alumina paste which is then fired; the wax burns out leaving an accurate mould from which one blade can be made by pouring liquid super-alloy into it (Fig. 20.4). Because the blades have to be made by this one-off method, they are expensive. One blade costs about UK£150 or US$330, of which only UK£10 (US$22) is materials; the total cost of a rotor of 102 blades is UK£15,300 or US$33,600.

Cast in this way, the grain size of such a blade is small (Fig. 20.4). The strengthening caused by alloying successfully suppresses power-law creep, but at $0.72T_M$, diffusional flow then becomes a problem (see the deformation-mechanism diagram of Chapter 19). The way out is to increase the grain size, or even make blades with no grain boundaries at all. In addition, *creep damage* (Chapter 17) accumulates at grain boundaries; we can

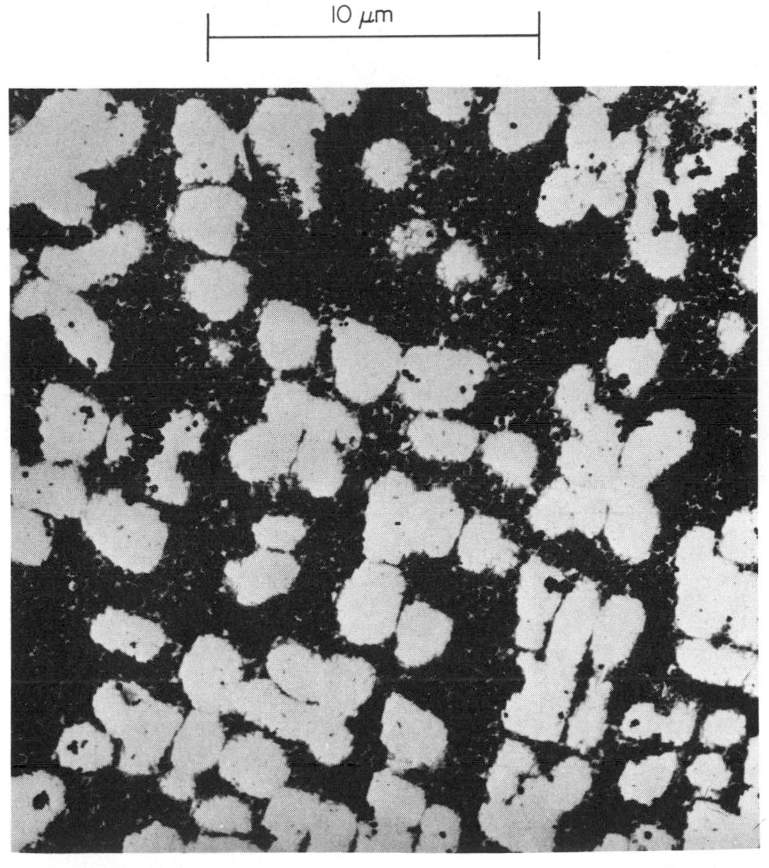

Fig. 20.3(a). A piece of a nickel-based super-alloy cut open to show the structure: there are two sizes of precipitate in the alloy—the large white precipitates, and the much smaller black precipitates in between.

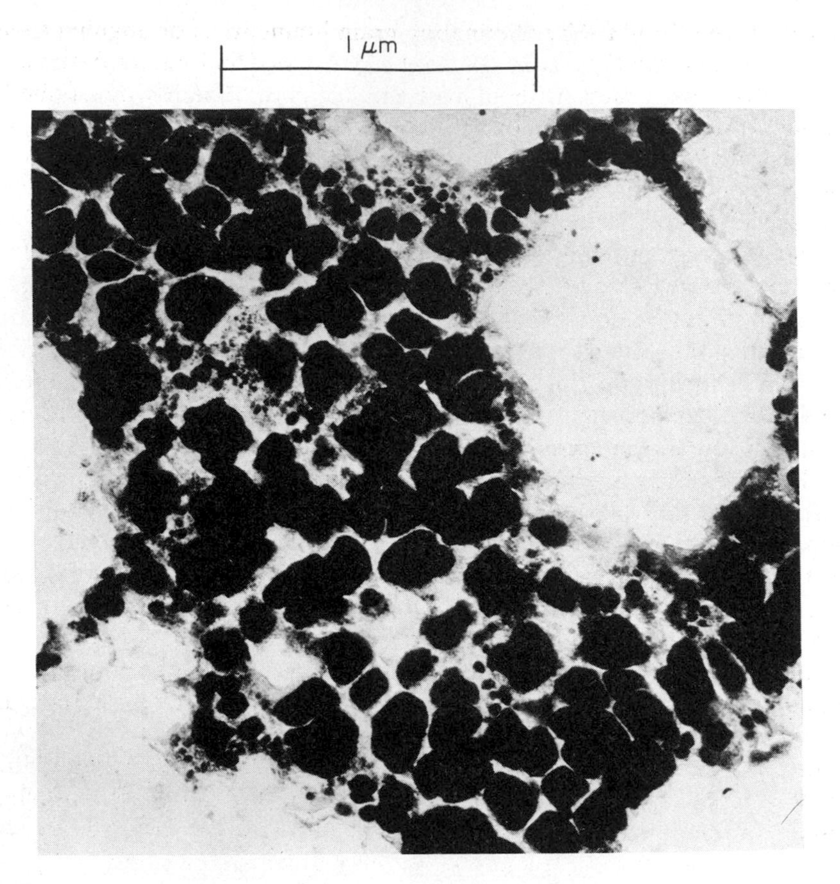

Fig. 20.3(b). As Fig. 20.3(a), but showing a much more magnified view of the structure, in which the small precipitates are more clearly identifiable.

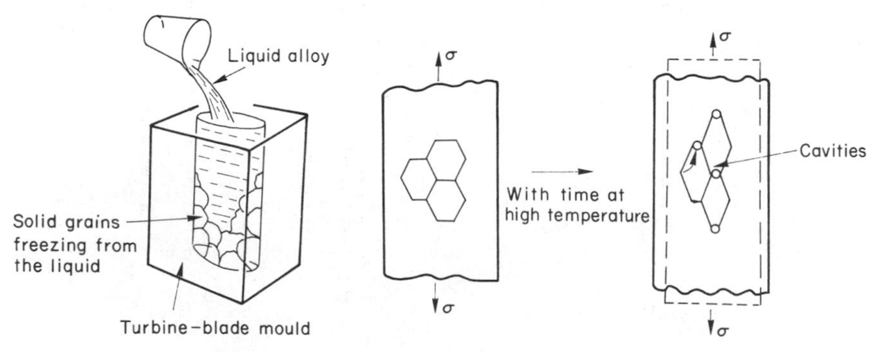

Fig. 20.4. Investment casting of turbine blades. This produces a fine-grained material which may undergo a fair amount of diffusion creep, and which may fail rather soon by cavity formation.

obviously stave off failure by eliminating grain boundaries, or aligning them parallel to the applied stress (see Fig. 20.4). To do this, we *directionally solidify* the alloys (see Fig. 20.5) to give long grains with grain boundaries parallel to the applied stress. The diffusional distances required for diffusional creep are then very large (greatly cutting down the rate of diffusional creep); in addition, there is no driving force for grain boundary sliding or for cavitation at grain boundaries. Directionally solidified alloys are under evaluation for creep properties at the moment, and will be in use in civil aircraft in two years or so. The improved creep properties of the DS alloy will allow the engine to run at a flame temperature approximately 50°C higher than at present, for an additional production cost of about UK£140 or US$310 per blade—the blades on one rotor will then cost UK£30,000 or US$66,000.

How was this type of alloy discovered in the first place? Well, the fundamental principles of creep-resistant materials design that we talked about helps us to select the more promising alloy recipes and discard the less promising ones fairly easily. There after, the approach is an empirical one. Large numbers of alloys having different recipes are made up in the laboratory, and tested for creep, oxidation, toughness, thermal fatigue and stability. The choice eventually narrows down to a few alloys and these are subjected to more stringent testing, coupled with judicious tinkering with the alloy recipe. All this is done using a semi-intuitive approach based on previous experience, knowledge of the basic principles of materials design and a certain degree of hunch and luck! Small improvements are continually made in alloy composition and in the manufacture of the finished blades, which evolve by a sort of creepy Darwinism, the fittest (in the sense of Table 20.1) surviving.

Figure 20.6 shows how this evolutionary process has resulted in a continual improvement of creep properties of nickel alloys over the last 30 years, and shows how the amounts of the major foreign elements have been juggled to obtain these improvements—keeping a watchful eye on the remaining necessary properties. The figure also shows how improvements in alloy manufacture—in this case the use of directional solidification—have helped to increase the operating temperature of the alloys. Nevertheless, it is clear from the graph that improvements in nickel alloys are now showing rapidly diminishing returns.

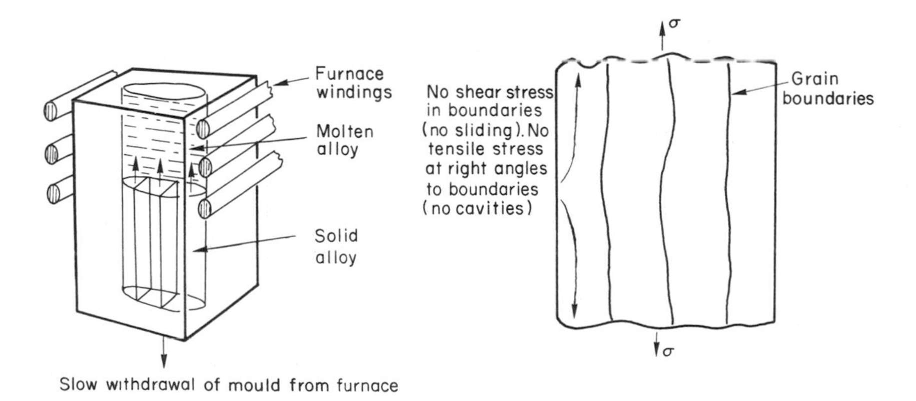

Fig. 20.5. Directional solidification (DS) of turbine blades. This gives blades with elongated grains, or even single-crystal blades with no grain boundaries at all.

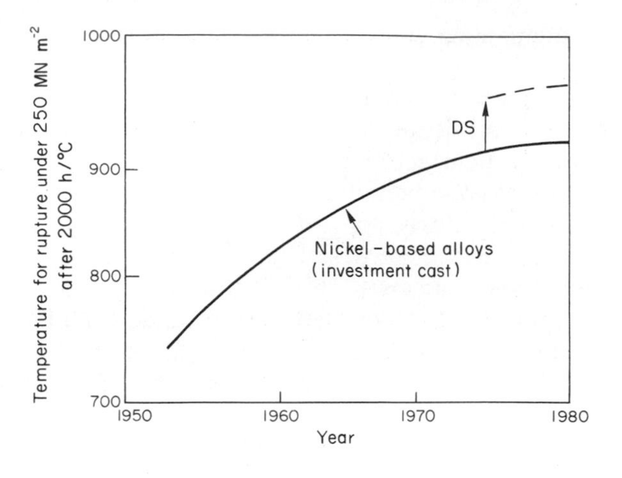

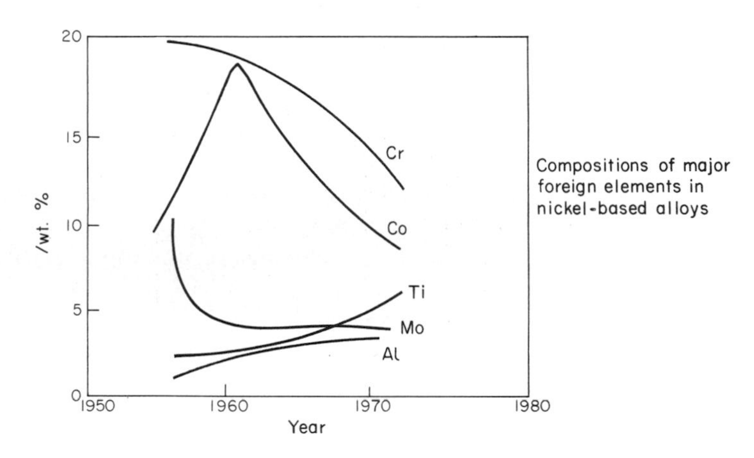

Fig. 20.6. Super-alloy developments.

Engineering developments—blade cooling

Figure 20.7 shows that up to 1960 turbine inlet temperatures were virtually the same as the metal temperatures. After 1960 there was a sharp divergence, with inlet temperatures substantially above the temperatures of the blade metal itself. This ability to run engines at a much higher temperature than would seem possible on the basis of the properties of the metal was due to the advent of air-cooled blades. In the earliest form of cooled blade, cooling air from the compressor stage of the engine was fed through ports passing along the full length of the blade, and was ejected into the gas stream at the end of the blade (see Fig. 20.8). This *internal cooling* of the blade enabled the inlet temperature to be increased immediately by 100°C with no improvement in alloy properties. A later improvement was *film cooling,* in which the air was ejected over the surface of the blade, giving a cool boundary layer between the blade and the hot gases. A programme of continuous improvement in the efficiency of heat transfer

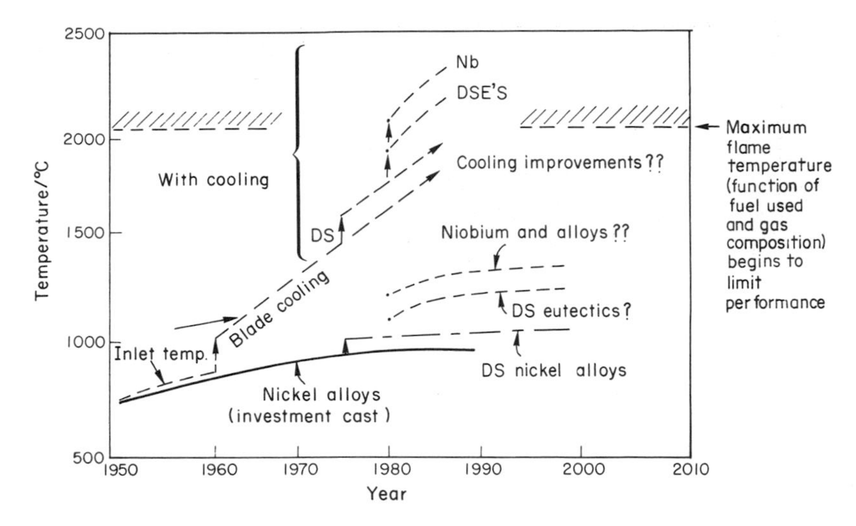

Fig. 20.7. Temperature evolution and future materials trends in turbine blades.

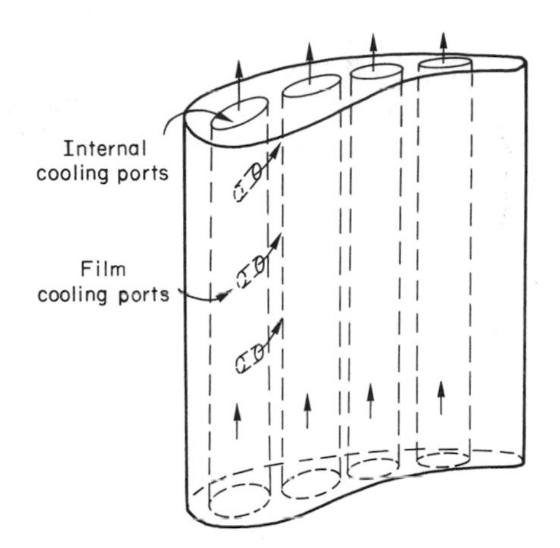

Fig. 20.8. Air-cooled blades.

by refinements of this type has made it possible for modern turbofans to operate at temperatures that are no longer dominated by the properties of the material.

Blade cooling has now, reached a limit: ducting still more cold air through the blades will begin to *reduce* the thermal efficiency by taking too much heat away from the combustion chamber. What, therefore, do we do now?

Future developments: metals and composites

In view of these impending limitations of blade cooling at its present stage of development, emphasis has now switched back to materials development. Because of the saturation in the development of nickel alloys, more revolutionary approaches are being tried.

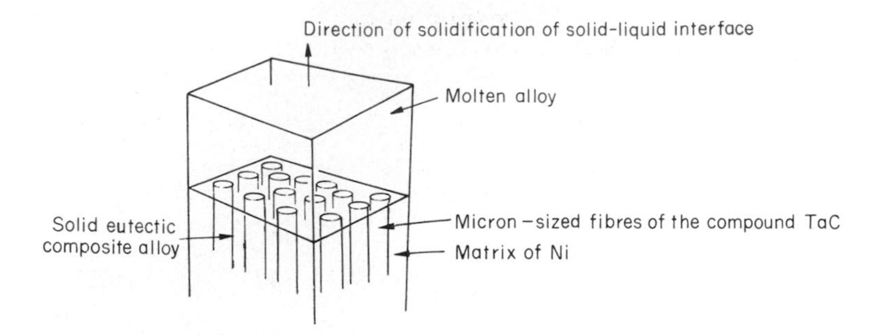

Fig. 20.9. Directionally solidified eutectics for turbine blades.

The major of these currently is to exploit a unique characteristic of some alloys—belonging to a class called *eutectic* alloys—to form spontaneously an aligned reinforced structure when directionally solidified. Figure 20.9 shows how they are made. Table 20.3 lists some typical high-temperature composites under study for turbine-blade applications. The reinforcing phase is usually a compound with a high melting temperature; this compound would be far too brittle on its own, but the surrounding metal matrix gives the composite the required toughness. The important feature of these alloys is that the high-melting-point fibres greatly improve the creep properties of the composite alloy, as one might anticipate. The bonding between fibres and matrix is atomic and therefore excellent. The structure is very fine (the fibres are only microns in diameter) and the accidental breaking of a few of the brittle fibres in service would have little effect on the composite as a whole.

As Fig. 20.7 shows, if DS eutectics prove successful, they will allow the metal temperature to be increased by ≈100°C above conventional DS nickel alloys, and the inlet temperature by ≈200°C (because of a temperature scaling effect caused by the blade cooling). Further improvements in alloy design are under way in which existing nickel alloys and DS eutectics are being blended to give a fibre-reinforced structure with precipitates in the matrix.

TABLE 20.3
HIGH-TEMPERATURE COMPOSITES

Matrix	Reinforcing phase	Reinforcing phase geometry
Ni	TaC	fibres
Co	TaC	fibres
Ni_3Al	Ni_3Nb	plates
Co	Cr_7C_3	fibres
Nb	Nb_2C	fibres

Future developments: high-temperature ceramics

The ceramics best suited for structural use at high temperatures (≳1000°C) are listed in Table 20.4 and compared with nickel-based super-alloys. The comparison shows that

TABLE 20.4
CERAMICS FOR HIGH-TEMPERATURE STRUCTURES

Material	Density /Mg m^{-3}	Melting or decomposition (D) temperature /K	Modulus /GN m^{-2}	Expansion coefficient $\times 10^{+6}$/K^{-1}	Thermal conductivity at 1000 K /W m^{-1} K^{-1}	Fracture toughness K_c/MN m$^{-3/2}$
Alumina, Al_2O_3	4.0	2320	360	6.9	7	≈5
Glass-ceramics (pyrocerams)	2.7	>1700	≈120	≈3	≈3	≈3
Hot-pressed silicon nitride, Si_3N_4	3.1	2173 (D)	310	3.1	16	≈5
Hot-pressed silicon carbide, SiC	3.2	3000 (D)	≈420	4.3	60	≈3.5
Nickel alloys (Nimonics)	8.0	1600	200	12.5	12	≈100

all the ceramics have attractively low densities, high moduli and high melting points (and thus excellent creep strength at 1000°C). But some have poor thermal conductivity (leading to high thermal stresses) and all have very low toughness.

Alumina (Al_2O_3) was one of the first pure oxides to be produced in complex shapes, but its combination of high expansion coefficient and poor conductivity and toughness gives it bad thermal-shock resistance.

Glass-ceramics are made by forming a complex silicate glass and then causing it to partly crystallise. They are widely used for ovenware and for heat exchangers for small engines. Their low thermal expansion gives them much better thermal shock resistance than most other ceramics, but the upper working temperature of 900°C (when the glass phase softens) makes them of limited use.

The covalently-bonded silicon carbide, silicon nitride, and sialons (alloys of Si_3N_4 and Al_2O_3) seem to be the best bet for high-temperature structural use. Their creep resistance is outstanding up to 1300°C, and their low expansion and high conductivity (better than nickel alloys!) makes them resist thermal shock well in spite of their typically low toughness. They can be formed by hot-pressing fine powders, or by nitriding silicon which is already pressed to shape: either way, precise shapes (like turbine blades) can be formed without the need for machining (they are much too hard to machine).

These materials are now under intensive study for turbine use.

Cost effectiveness

Any major materials development programme, such as that on the eutectic super-alloys, can only be undertaken if a successful outcome would be cost effective. As Fig. 20.10 shows, the costs of development can be colossal. Even before a new material is out of the laboratory, 2 to 4 million pounds (4 to 8 million dollars) can have been spent, and failure in an engine test can be expensive. Because the performance of a new alloy cannot finally be verified until it has been in flight, at each stage of development risk decisions have to be taken whether to press ahead, or cut losses and abandon the programme. Such decisions are currently being taken about eutectic super-alloys and in all probability, considerable further small-scale research on these systems will be

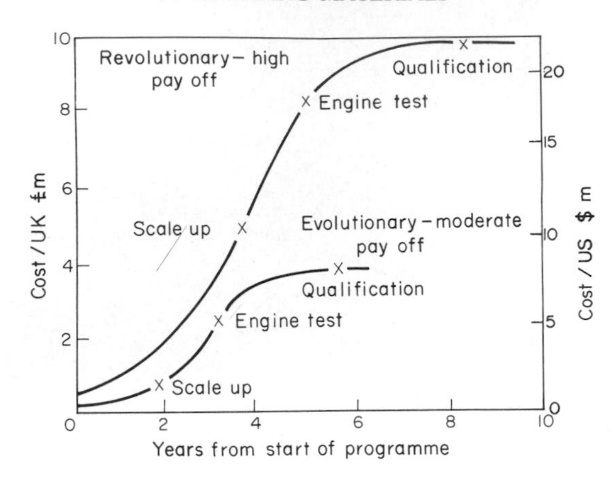

Fig. 20.10. Development costs of new turbine-blade materials.

required before they can be sanctioned for scale-up testing. One must consider, too, the the cost of the materials themselves. Some of the metals used in conventional nickel alloys—such as hafnium—are hideously expensive (at UK£100,00 tonne^{-1} or US$220,000 tonne^{-1}) and extremely scarce; and the use of greater and greater quantities of exotic materials in an attempt to improve the creep properties will drive the cost of blades up. But expensive though it is, the HP1 set costing UK£30,000 (US$66,000) is still only a small fraction of the cost of an engine or of the fuel it will consume. Blade costs *are* important, but if new alloys offer improved life or inlet temperature, there is a strong incentive to pursue them.

F. Oxidation and corrosion

CHAPTER 21

OXIDATION OF MATERIALS

Introduction

In the last chapter we said that one of the requirements of a high-temperature material—in a turbine blade, or a super-heater tube, for example—was that it should resist attack by gases at high temperatures and, in particular, that it should resist oxidation. Turbine blades *do* oxidise in service, and react with H_2S, SO_2 and other combustion products. Excessive attack of this sort is obviously undesirable in such a highly stressed component. Which materials best resist oxidation, and how can the resistance to gas attack be improved?

Well, the earth's atmosphere is oxidising. We can get some idea of oxidation-resistance by using the earth as a laboratory, and looking for materials which survive well in its atmosphere. All around us we see ceramics: the earth's crust (Chapter 2) is almost entirely made of oxides, silicates, aluminates and other compounds of oxygen; and being oxides already, they are completely stable. Alkali halides, too, are stable: NaCl, KCl, NaBr—all are widely found in nature. By contrast, metals are not stable: only gold is found in "native" form under normal circumstances (it is completely resistant to oxidation at all temperatures); all the others in our data sheets will oxidise in contact with air. Polymers are not stable either: most will burn if ignited, meaning that they oxidise readily. Coal and oil (the raw materials for polymers), it is true, are found in nature, but that is only because geological accidents have sealed them off from all contact with air. A few polymers, among them PTFE (a polymer based on —CF_2—), are so stable that they survive long periods at high temperatures, but they are the exceptions. And polymer-based composites, of course, are just the same: wood is not noted for its high-temperature oxidation resistance.

How can we categorise in a more precise way the oxidation-resistance of materials? If we can do so for oxidation, we can obviously follow a similar method for sulphidation or nitrogenation.

The energy of oxidation

This tendency of many materials to react with oxygen can be quantified by laboratory tests which measure the energy needed for the reaction

$$\text{Material} + \text{Oxygen} + \text{Energy} \rightarrow \text{Oxide of material.}$$

If this energy is *positive*, the material is stable; if *negative*, it will oxidise. The bar-chart of Fig. 21.1 shows the energies of oxide formation for our four categories of materials; numerical values are given in Table 21.1.

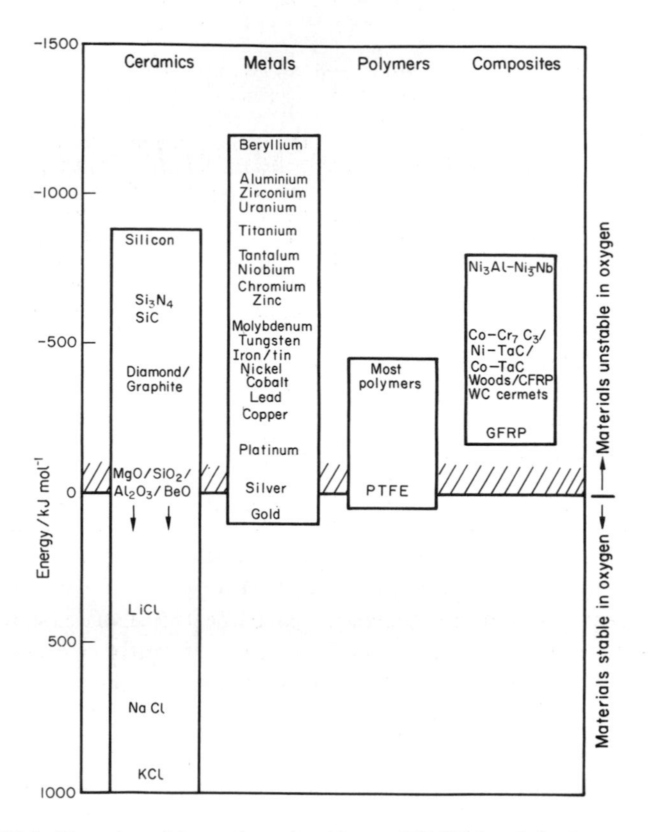

Fig. 21.1. Energies of formation of oxides at 273 K/kJ mol^{-1} of oxygen O_2.

Rates of oxidation

When designing with oxidation-prone materials, it is obviously vital to know how *fast* the oxidation process is going to be. Intuitively one might expect that, the larger the energy released in the oxidation process, the faster the *rate* of oxidation. For example, one might expect aluminium to oxidise 2.5 times faster than iron from the energy data in Fig. 21.1. In fact, aluminium oxidises much more slowly than iron. Why should this happen?

If you heat a piece of bright iron in a gas flame, the oxygen in the air reacts with the iron at the surface of the metal where the oxygen and iron atoms can contact, creating a thin layer of iron oxide on the surface, and making the iron turn black. The layer grows in thickness, quickly at first, and then more slowly because iron atoms now have to diffuse through the film before they make contact and react with oxygen. If you plunge the piece of hot iron into a dish of water the shock of the quenching breaks off the iron

TABLE 21.1
ENERGIES OF FORMATION OF OXIDES AT 273 K

Material (oxide)		Energy /kJ mol^{-1} of oxygen, O_2	Material (oxide)		Energy /kJ mol^{-1} of oxygen, O_2
Beryllium	(BeO)	−1182	Cobalt	(CoO)	−422
Magnesium	(MgO)	−1162	Woods, most polymers, CFRP		≈−400
Aluminium	(Al_2O_3)	−1045	Diamond, graphite	(CO_2)	−389
Zirconium	(ZrO_2)	−1028	Tungsten carbide cermet (mainly WC)	(WO_3 + CO_2)	−349
Uranium	(U_3O_8)	≈ − 1000	Lead	(Pb_3O_4)	−309
Titanium	(TiO)	−848	Copper	(CuO)	−254
Silicon	(SiO_2)	−836	GFRP		≈ − 200
Tantalum	(Ta_2O_5)	−764	Platinum	(PtO_2)	≈ − 160
Niobium	(Nb_2O_5)	−757	Silver	(Ag_2O)	−5
Chromium	(Cr_2O_3)	−701	PTFE		≈zero
Zinc	(ZnO)	−636	Gold	(Au_2O_3)	+80
Silicon nitride Si_3N_4	($3SiO_2$ + $2N_2$)	≈−629	Alkali halides		≈ +400 to ≈ + 1400
Silicon carbide SiC	(SiO_2 + CO_2)	≈−580			
Molybdenum	(MoO_2)	−534	Magnesia, MgO	Higher oxides	Large and positive
Tungsten	(WO_3)	−510	Silica, SiO_2		
Iron	(Fe_3O_4)	−508	Alumina, Al_2O_3		
Tin	(SnO)	−500	Beryllia, BeO		
Nickel	(NiO)	−439			

oxide layer, and you can see the pieces of layer in the dish. The iron surface now appears bright again, showing that the shock of the quenching has completely stripped the metal of the oxide layer which formed during the heating; if it were reheated, it would oxidise at the old rate.

The important thing about the oxide film is that it acts as a *barrier* which keeps the oxygen and iron atoms apart and cuts down the rate at which these atoms react to form more iron oxide. Aluminium, and most other materials, form oxide barrier layers in just the same sort of way—but the oxide layer on aluminium is a *much* more effective barrier than the oxide film on iron is.

How do we measure rates of oxidation in practice? Well, because oxidation proceeds by the addition of oxygen atoms to the surface of the material, the *weight* of the material usually goes up in proportion to the amount of material that has become oxidised. This weight increase, Δm, can be monitored continuously with time t in the way illustrated in Fig. 21.2. Two types of behaviour are usually observed at high temperature. The first is *linear oxidation*, with

$$\Delta m = k_L t \qquad (21.1)$$

where k_L is a *kinetic constant*. Naturally, k_L is usually positive. (In a few materials, however, the oxide evaporates away as soon as it has formed; the material then *loses* weight and k_L is then negative.)

The second type of oxidation behaviour is *parabolic oxidation*, with

$$(\Delta m)^2 = k_P t \qquad (21.2)$$

where k_P is another kinetic constant, this time always positive.

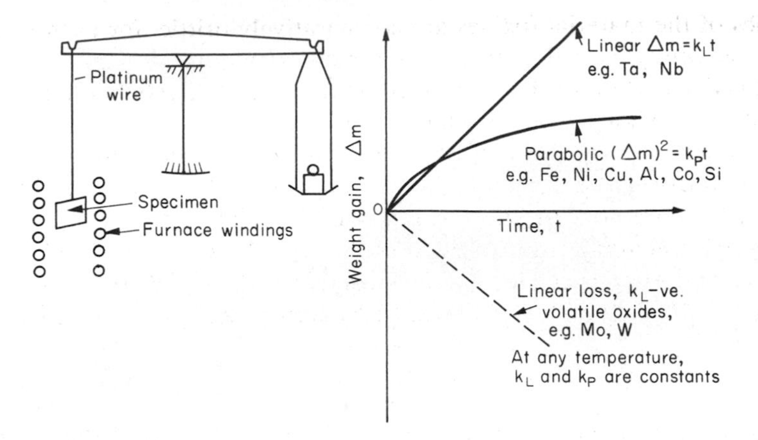

Fig. 21.2. Measurement of oxidation rates.

Oxidation rates follow Arrhenius's Law (Chapter 18), that is, the kinetic constants k_L and k_P increase exponentially with temperature:

$$k_L = A_L e^{-Q_L/\bar{R}T} \quad \text{and} \quad k_P = A_P e^{-Q_P/\bar{R}T} \tag{21.3}$$

where A_L and A_P, Q_L and Q_P are constants. Thus, as the temperature is increased, the rate of oxidation increases exponentially (Fig. 21.3).

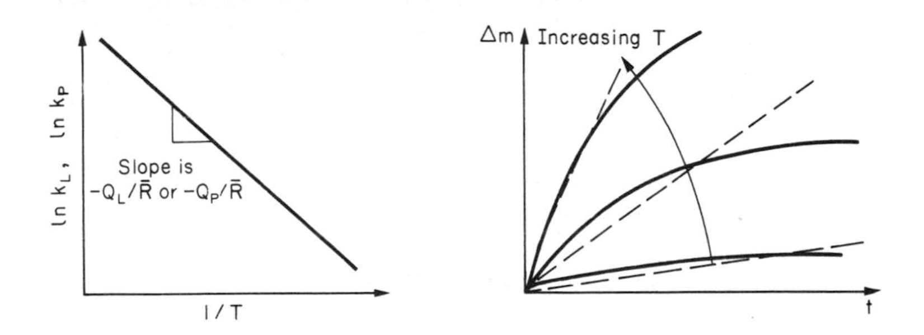

Fig. 21.3. Oxidation rates increase with temperature according to Arrhenius's Law.

Finally, oxidation rates obviously increase with increasing partial pressure of oxygen, although rarely in a simple way. The partial pressure of oxygen in a gas turbine atmosphere, for example, may well be very different from that in air, and it is important to conduct oxidation tests on high-temperature components under the right conditions.

Data

Naturally, it is important from the design standpoint to know how much material is replaced by oxide. The mechanical properties of the oxide are usually grossly inferior to the properties of the material (oxides are comparatively brittle, for example), and even if the layer is firmly attached to the material—which is certainly not always the case—the effective section of the component is reduced. The reduction in the section of a component can obviously be calculated from data for Δm.

Table 21.2 gives the times for a range of materials required to oxidise them to a depth of 0.1 mm from the surface when exposed to air at $0.7T_M$ (a typical figure for the operating temperature of a turbine blade or similar component): these times vary by many orders of magnitude, and clearly show that there is no correlation between oxidation rate and energy needed for the reaction (see Al, W for extremes: Al, *very* slow—energy $= -1045$ kJ mol^{-1} of O_2; W, very fast—energy $= -510$ kJ mol^{-1} of O_2).

TABLE 21.2

Time in hours for material to be oxidised to a depth of 0.1 mm at 0.7 T_M in air (N.B.—data subject to considerable variability due to varying degrees of material purity, prior surface treatment and presence of atmospheric impurities like sulphur)

Material	Time	Melting point/K	Material	Time	Melting point/K
Au	infinite?	1336	Ni	600	1726
Ag	very long	1234	Cu	25	1356
Al	very long	933	Fe	24	1809
Si_3N_4	very long	2173	Co	7	1765
SiC	very long	3110	Ti	<6	1943
Sn	very long	505	WC cermet	<5	1700
Si	2×10^6	1683	Ba	≪0.5	983
Be	10^6	1557	Zr	0.2	2125
Pt	1.8×10^5	2042	Ta	very short	3250
Mg	$>10^5$	923	Nb	very short	2740
Zn	$>10^4$	692	U	very short	1405
Cr	1600	2148	Mo	very short	2880
Na	>1000	371	W	very short	3680
K	>1000	337			

Micromechanisms

Figure 21.4 illustrates the mechanism of parabolic oxidation. The reaction

$$M + O \rightarrow MO$$

(where M is the material which is oxidising and O is oxygen) really goes in two steps. First M forms an ion, releasing electrons, e:

$$M \rightarrow M^{++} + 2e.$$

These electrons are then absorbed by oxygen to give an oxygen ion:

$$O + 2e \rightarrow O^{--}.$$

Either the M^{++} and the two e's diffuse outward through the film to meet the O^{--} at the outer surface, *or* the oxygen diffuses inwards (with two electron holes) to meet the

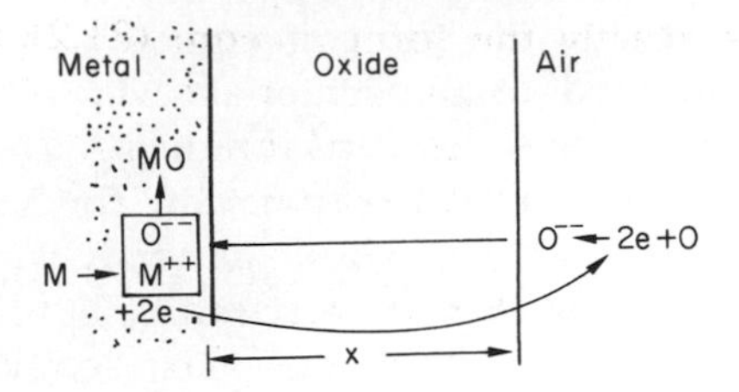

Case 1. M^{++} diffuses very slowly in oxide. Oxide grows at metal–oxide interface. Examples: Ti, Zr, U.

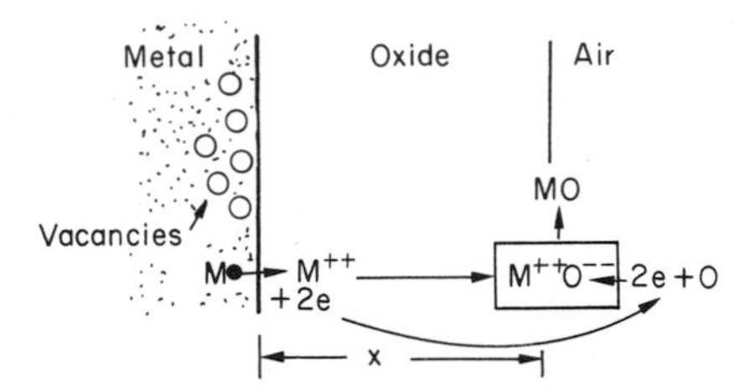

Case 2. O^{--} diffuses very slowly in oxide. Oxide grows at oxide–air interface. Vacancies form between metal and oxide. Examples: Cu, Fe, Cr, Co.

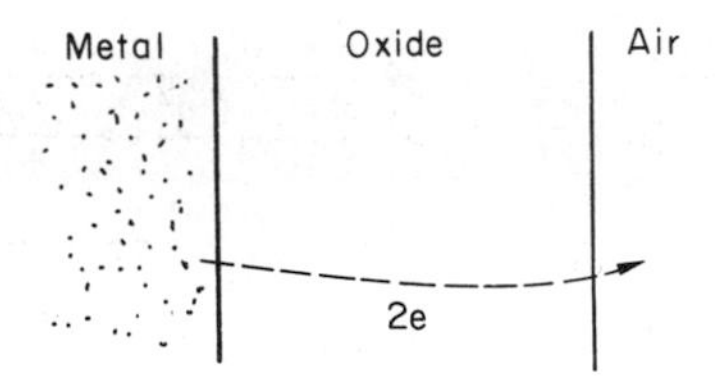

Case 3. Electrons move very slowly. Oxide can grow (slowly) at metal–oxide interface or oxide–air interface depending on whether M^{++} diffuses faster than O^{--} or not. Example: Al.

Fig. 21.4. How oxide layers grow to give parabolic oxidation behaviour.

M^{++} at the inner surface. The concentration gradient of oxygen is simply the concentration in the gas, c, divided by the film thickness, x; and the rate of growth of the film $\mathrm{d}x/\mathrm{d}t$ is obviously proportional to the flux of atoms diffusing through the film. So, from Fick's Law (eqn. (18.1)):

$$\frac{\mathrm{d}x}{\mathrm{d}t} \propto D\frac{c}{x}$$

where D is the diffusion coefficient. Integrating with respect to time gives

$$x^2 = K_p t \tag{21.4}$$

where

$$K_p \propto cD_0 e^{-Q/\bar{R}T}. \tag{21.5}$$

This growth law has exactly the form of eqn. (21.2) and the kinetic constant is analogous to* that of eqn. (21.3). This success lets us explain why some films are more protective than others: protective films are those with low diffusion coefficients—and thus high melting points. That is one reason why Al_2O_3 protects aluminium, Cr_2O_3 protects chromium and SiO_2 protects silicon so well, whereas Cu_2O and even FeO (which have lower melting points) are less protective. But there is an additional reason: electrons must also pass through the film and these films are insulators (the electrical resistivity of Al_2O_3 is 10^9 times greater than that of FeO).

Although our simple oxide film model explains most of the experimental observations we have mentioned, it does not explain the linear laws. How, for example, can a material *lose* weight linearly when it oxidises as is sometimes observed (see Fig. 21.2)? Well, some oxides (e.g. MoO_3, WO_3) are very volatile. During oxidation of Mo and W at high temperature, the oxides evaporate as soon as they are formed, and offer no barrier at all to oxidation. Oxidation, therefore, proceeds at a rate that is independent of time, and the material loses weight because the oxide is *lost*. This behaviour explains the catastrophically rapid section loss of Mo and W shown in Table 21.2.

The explanation of a linear weight *gain* is more complex. Basically, as the oxide film thickens, it develops cracks, or partly lifts away from the material, so that the barrier between material and oxide does not become any more effective as oxidation proceeds. Figure 21.5 shows how this can happen. If the volume of the oxide is much less than

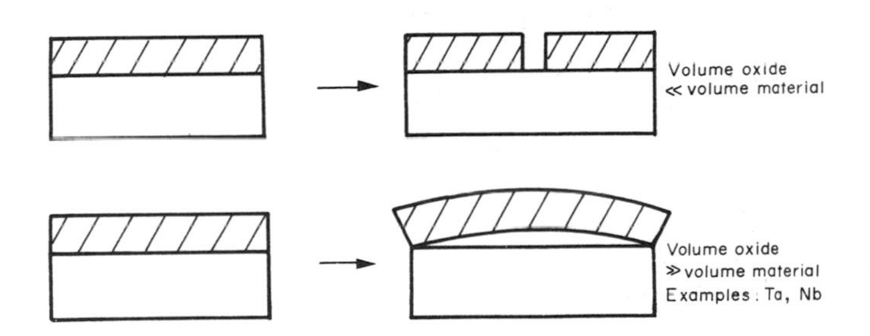

Fig. 21.5. Breakdown of oxide films, leading to linear oxidation behaviour.

that of the material from which it is formed, it will crack to relieve the strain (oxide films are usually brittle). If the volume of the oxide is much greater, on the other hand, the oxide will tend to release the strain energy by breaking the adhesion between material and oxide, and springing away. For protection, then, we need an oxide skin which is neither too small and splits open (like the bark on a fir tree) nor one which is too big and wrinkles up (like the skin of a rhinoceros), but one which is just right. Then, and only then, do we get protective parabolic growth.

In the next chapter we use this understanding to analyse the design of oxidation-resistant materials.

* It does not have the same value, however, because eqn. (21.5) refers to thickness gain and not *mass* gain; the two can be easily related if quantities like the density of the oxide are known.

Further reading

J. P. Chilton, *Principles of Metallic Corrosion*, 2nd edition, The Chemical Society, 1973, Chap. 2.
M. G. Fontana and N. D. Greene, *Corrosion Engineering*, McGraw Hill, 1967, Chap. 11.
J. C. Scully, *The Fundamentals of Corrosion*, 2nd edition, Pergamon Press, 1975, Chap. 1.
O. Kubaschewski and B. E. Hopkins, *Oxidation of Metals and Alloys*, 2nd edition, Butterworths, 1962.
C. J. Smithells, *Metals Reference Book*, 5th edition, Butterworths, 1976 (for data).

CHAPTER 22

CASE STUDIES IN DRY OXIDATION

Introduction

In this chapter we look first at an important class of alloys designed to resist corrosion: the stainless steels. We then examine a more complicated problem: that of protecting the most advanced gas turbine blades from gas attack. The basic principle applicable to both cases is to coat the steel or the blade with a stable ceramic: usually Cr_2O_3 or Al_2O_3. But the ways this is done differ widely. The most successful are those which produce a ceramic film which heals itself if damaged—as we shall now describe.

Case study 1: making stainless alloys

Mild steel is an excellent structural material—cheap, easily formed and strong mechanically. But at low temperatures it rusts, and at high, it oxidises rapidly. There is a demand, for applications ranging from kitchen sinks via chemical reactors to superheater tubes, for a corrosion-resistant steel. In response to this demand, a range of stainless irons and steels has been developed. When mild steel is exposed to hot air, it oxidises quickly to form FeO (or higher oxides). But if one of the elements near the top of Table 21.1 with a large energy of oxidation is dissolved in the steel, then this element oxidises preferentially (because it is much more stable than FeO), forming a layer of its oxide on the surface. And if this oxide is a protective one, like Cr_2O_3, Al_2O_3, SiO_2 or BeO, it stifles further growth, and protects the steel.

A considerable quantity of this foreign element is needed to give adequate protection. The best is chromium, 18% of which gives a very protective oxide film: it cuts down the rate of attack at 900°C, for instance, by more than 100 times.

Other elements, when dissolved in steel, cut down the rate of oxidation, too. Al_2O_3 and SiO_2 both form in preference to FeO (Table 21.1) and form protective films (see Table 21.2). Thus 5% Al dissolved in steel decreases the oxidation rate by 30 times, and 5% Si by 20 times. The same principle can be used to impart corrosion resistance to other metals. We shall discuss nickel and cobalt in the next case study—they can be alloyed in this way. So, too, can copper; although it will not dissolve enough chromium to give a good Cr_2O_3 film, it *will* dissolve enough aluminium, giving a range of stainless alloys called "aluminium bronzes". Even silver can be prevented from tarnishing (reaction with sulphur) by alloying it with aluminium or silicon, giving protective Al_2O_3 or SiO_2 surface films. And archaeologists believe that the Delhi Pillar—an ornamental pillar of cast iron which has stood, uncorroded, for some hundreds of years in a *particularly* humid spot—survives because the iron has some 6% silicon in it.

Ceramics themselves are sometimes protected in this way. Silicon carbide, SiC, and silicon nitride, Si_3N_4, both have large negative energies of oxidation (meaning that they oxidise easily). But when they do, the silicon in them turns to SiO_2 which quickly forms a protective skin and prevents further attack.

This protection-by-alloying has one great advantage over protection by a surface coating (like chromium plating or gold plating): it repairs itself when damaged. If the protective film is scored or abraded, fresh metal is exposed, and the chromium (or aluminium or silicon) it contains immediately oxidises, healing the break in the film.

Case study 2: Protecting turbine blades

As we saw in Chapter 20, the materials at present used for turbine blades consist chiefly of nickel, with various foreign elements added to get the creep properties right. With the advent of DS blades, such alloys will normally operate around 950°C, which is close to $0.7T_M$ for Ni (1208 K, 935°C). If we look at Table 21.2 we can see that at this temperature, nickel loses 0.1 mm of metal from its surface by oxidation in 600 hours. Now, the thickness of the metal between the outside of the blade and the integral cooling ports is about 1 mm, so that in 600 hours a blade would lose about 10% of its cross-section in service. This represents a serious loss in mechanical integrity and, moreover, makes no allowance for statistical variations in oxidation rate—which can be quite large—or for preferential oxidation (at grain boundaries, for example) leading to pitting. Because of the large cost of replacing a set of blades (≈UK£15,000 or US$33,000 per engine) they are expected to last for more than 5000 hours. Nickel oxidises with parabolic kinetics (eqn. (21.4)) so that, after a time t_2, the loss in section x_2 is given by substituting our data into:

$$\frac{x_2}{x_1} = \left(\frac{t_2}{t_1}\right)^{1/2}$$

giving

$$x_2 = 0.1\left(\frac{5000}{600}\right)^{1/2} = 0.29\ \text{mm}.$$

Obviously this sort of loss is not admissible, but how do we stop it?

Well, as we saw in Chapter 20, the alloys used for turbine blades contain large amounts of chromium, dissolved in solid solution in the nickel matrix. Now, if we look at our table of energies (Table 21.1) released when oxides are formed from materials, we see that the formation of Cr_2O_3 releases much more energy (701 kJ mol^{-1} of O_2) than NiO (439 kJ mol^{-1} of O_2). This means that Cr_2O_3 will form *in preference* to NiO on the surface of the alloy. Obviously, the more Cr there is in the alloy, the greater is the preference for Cr_2O_3. At the 20% level, enough Cr_2O_3 forms on the surface of the turbine blade to make the material act a bit as though it were chromium.

Suppose for a moment that our material *is* chromium. Table 21.2 shows that Cr would lose 0.1 mm in 1600 hours at $0.7T_M$. Of course, we have forgotten about one thing. $0.7T_M$ for Cr is 1504 K (1231°C), whereas, as we have said, for Ni, it is 1208 K

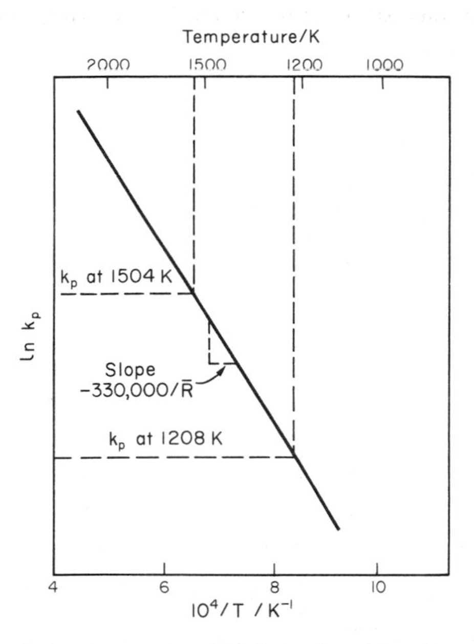

Fig. 22.1. The way in which k_p varies with temperature.

(935°C). We should, therefore, consider how Cr_2O_3 would act as a barrier to oxidation at 1208 K rather than at 1504 K (Fig. 22.1). The oxidation of chromium follows parabolic kinetics with an activation energy of 330 kJ mol^{-1}. Then the ratio of the times required to remove 0.1 mm (from eqn. (21.3)) is

$$\frac{t_2}{t_1} = \frac{\exp-(Q/\bar{R}T_1)}{\exp-(Q/\bar{R}T_2)} = 0.65 \times 10^3.$$

Thus the time at 1208 K is

$$t_2 = 0.65 \times 10^3 \times 1600 \text{ hours}$$
$$= 1.04 \times 10^6 \text{ hours.}$$

Now, as we have said, there is only at most 20% Cr in the alloy, and the alloy behaves only *partly* as if it were protected by Cr_2O_3. In fact, experimentally, we find that 20% Cr increases the time for a given metal loss by only about *ten* times, i.e. the time taken to lose 0.1 mm at blade working temperature becomes 600×10 hours = 6000 hours rather than 10^6 hours.

Why this large difference? Well, whenever you consider an *alloy* rather than a pure material, the oxide layer—whatever its nature (NiO, Cr_2O_3, etc.)—has foreign elements contained in *it*, too. Some of these will greatly increase either the diffusion coefficients in, or electrical conductivity of, the layer, and make the rate of oxidation through the layer much more than it would be in the absence of foreign element contamination. One therefore has to be very careful in transferring data on film protectiveness from a pure material to an alloyed one, but the approach does, nevertheless, give us an *idea* of what to expect. As in all oxidation work, however, experimental determinations on actual alloys are *essential* for working data.

This 0.1 mm loss in 6000 hours from a 20% Cr alloy at 935°C, though better than pure nickel, is still not good enough. What is worse, we saw in Chapter 20 that, to improve the creep properties, the quantity of Cr has been reduced to 10%, and the resulting oxide film is even less protective. The obvious way out of this problem is to *coat* the blades with a protective layer (Fig. 22.2). This is usually done by spraying

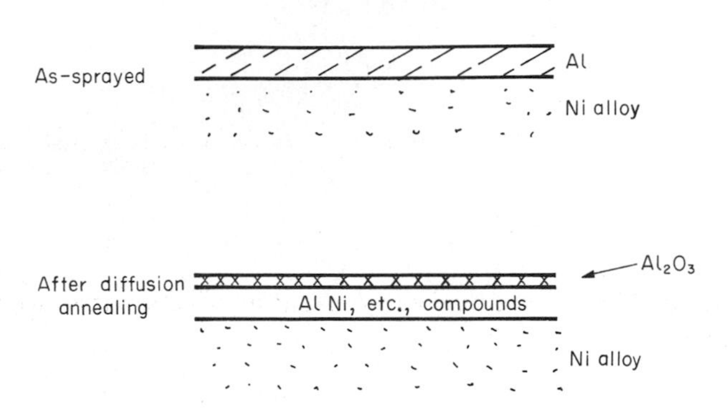

Fig. 22.2. Protection of turbine blades by sprayed-on aluminium.

molten droplets of aluminium on to the blade surface to form a layer, some microns thick. The blade is then heated in a furnace to allow the Al to diffuse into the surface of the Ni. During this process, some of the Al forms compounds such as AlNi with the nickel—which are themselves good barriers to oxidation of the Ni, whilst the rest of the Al becomes oxidised up to give Al_2O_3—which, as we can see from our oxidation-rate data—should be a very good barrier to oxidation even allowing for the high temperature ($0.7T_M$ for Al = 653 K, 380°C). An incidental benefit of the relatively thick AlNi layer is its poor thermal conductivity—this helps insulate the metal of the cooled blade from the hot gases, and allows a slight extra increase in blade working temperature.

Other coatings, though more difficult to apply, are even more attractive. AlNi is brittle, so there is a risk that it may chip off the blade surface exposing the unprotected metal. It is possible to diffusion-bond a layer of a Ni–Cr–Al alloy to the blade surface (by spraying on a powder or pressing on a thin sheet and then heating it up) to give a ductile coating which still forms a very protective film of oxide.

Influence of coatings on mechanical properties

So far, we have been talking in our case study about the *advantage* of an oxide layer in reducing the rate of metal removal by oxidation. Oxide films do, however, have some *disadvantages.*

Because oxides are usually quite brittle at the temperatures encountered on a turbine blade surface, they can crack, especially when the temperature of the blade changes and differential thermal contraction and expansion stresses are set up between alloy and oxide. These can act as ideal nucleation centres for thermal fatigue cracks and, because oxide layers in nickel alloys are stuck well to the underlying alloy (they would be useless if they were not), the crack can spread into the alloy itself (Fig. 22.3). The

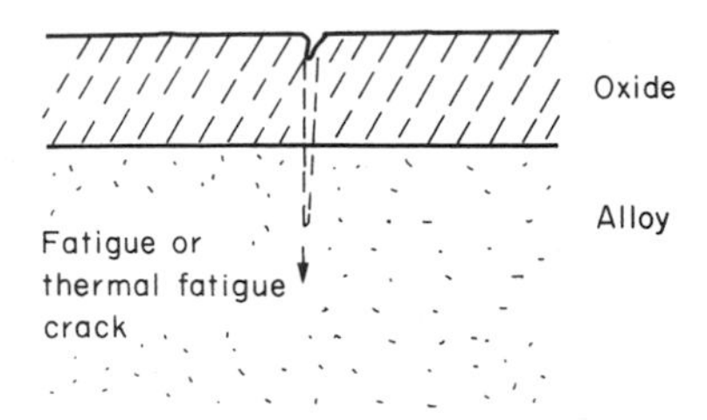

Fig. 22.3. Fatigue cracks can spread from coatings into the material itself.

properties of the oxide film are thus very important in affecting the fatigue properties of the whole component.

Protecting future blade materials

What of the corrosion resistance of new turbine-blade alloys like DS eutectics? Well, an alloy like Ni_3Al–Ni_3Nb loses 0.05 mm of metal from its surface in 48 hours at the anticipated operating temperature of 1155°C for such alloys. This is obviously not a good performance, and coatings will be required before these materials are suitable for application. At lower oxidation rates, a more insidious effect takes place—preferential attack of one of the phases, with penetration along interphase boundaries. Obviously this type of attack, occurring under a break in the coating, can easily lead to fatigue failure and raises another problem in the use of DS eutectics.

You may be wondering why we did not mention the pure "refractory" metals Nb, Ta, Mo, W in our chapter on turbine-blade materials (although we *did* show one of them on Fig. 20.7). These metals have very high melting temperatures, as shown, and should therefore have very good creep properties.

$$\left.\begin{array}{ll} \text{Nb} & 2740\text{ K} \\ \text{Ta} & 3250\text{ K} \\ \text{Mo} & 2880\text{ K} \\ \text{W} & 3680\text{ K} \end{array}\right\} T_M.$$

But they all oxidise very rapidly indeed (see Table 21.2), and are utterly useless without coatings. The problem with coated refractory metals is, that if a break occurs in the coating (e.g. by thermal fatigue, or erosion by dust particles, etc.), catastrophic oxidation of the underlying metal will take place, leading to rapid failure. The "unsafeness" of this situation is a major problem that has to be solved before we can use these on-other-counts potentially excellent materials.

The ceramics SiC and Si_3N_4 do not share this problem. They oxidise readily (Table 21.1); but in doing so, a surface film of SiO_2 forms which gives adequate protection up to 1300°C. And because the film forms by oxidation of the material itself, it is self-healing.

Joining operations: a final note

One might imagine that it is always a good thing to have a protective oxide film on a material. Not always; if you wish to join materials by brazing or soldering, the protective oxide film can be a problem. It is this which makes stainless steel hard to braze and almost impossible to solder; even spot-welding and diffusion bonding become difficult. Protective films create poor electrical contacts; that is why aluminium is not more widely used as a conductor. And production of components by powder methods (which involve the compaction and sintering—really diffusion bonding—of the powdered material to the desired shape) is made difficult by protective surface films.

Further reading

M. G. Fontana and N. D. Greene, *Corrosion Engineering*, McGraw Hill, 1967, Chap. 11.
D. R. Gabe, *Principles of Metal Surface Treatment and Protection*, 2nd edition, Pergamon Press, 1978.

CHAPTER 23

WET CORROSION OF MATERIALS

Introduction

In the last two chapters we showed that most materials that are unstable in oxygen tend to oxidise. We were principally concerned with loss of material at high temperatures, in dry environments, and found that, under these conditions, oxidation was usually controlled by the diffusion of ions or the conduction of electrons through oxide films that formed on the material surface (Fig. 23.1). Because of the thermally activated

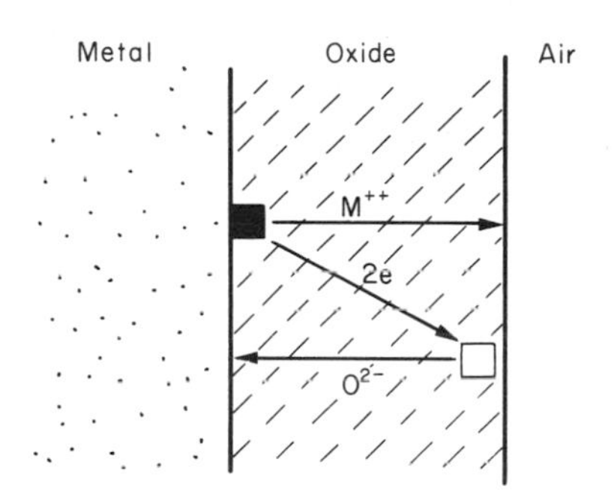

Fig. 23.1. *Dry* oxidation.

nature of the diffusion and reaction processes we saw that the rate of oxidation was much greater at high temperature than at low, although even at room temperature, very thin films of oxide *do* form on all unstable metals. This minute amount of oxidation is important: it protects, preventing further attack; it causes tarnishing; it makes joining difficult; and (as we shall see in Chapters 25 and 26) it helps keep sliding surfaces apart, and so influences the coefficient of friction. But the *loss* of material by oxidation at room temperature under these *dry* conditions is very slight.

Under *wet* conditions, the picture is dramatically changed. When mild steel is exposed to oxygen and water at room temperature, it rusts rapidly and the loss of metal quickly becomes appreciable. Unless special precautions are taken, the life of most structures, from bicycles to bridges, from buckets to battleships, is limited by wet corrosion. The annual bill in the UK either replacing corroded components, or preventing corrosion (e.g. by painting the Forth Bridge), is around UK£2000 m or US$4400 m a year.

Wet corrosion

Why the dramatic effect of water on the rate of loss of material? As an example we shall look at *iron*, immersed in *aerated water* (Fig. 23.2).

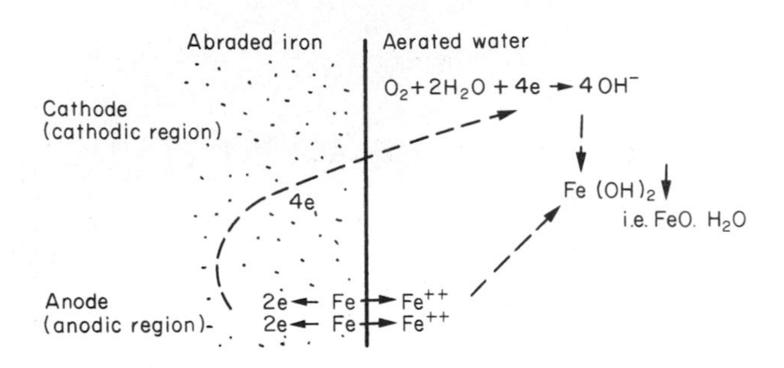

Fig. 23.2. *Wet* corrosion.

Iron atoms pass into solution in the water as Fe^{++}, leaving behind two electrons each (the *anodic* reaction). These are conducted through the metal to a place where the "oxygen reduction" reaction can take place to consume the electrons (the *cathodic* reaction). This reaction generates OH^- ions which then combine with the Fe^{++} ions to form a *hydrated iron oxide* $Fe(OH)_2$ (really FeO, H_2O); but instead of forming on the surface where it might give some protection, it often forms as a precipitate in the water itself. The reaction can be summarised by

$$\text{Material} + \text{Oxygen} \rightarrow \text{(Hydrated) Material Oxide}$$

just as in the case of dry oxidation.

Now the formation and solution of Fe^{++} is analogous to the formation and diffusion of M^{++} in an oxide film under dry oxidation; and the formation of OH^- is closely similar to the reduction of oxygen on the surface of an oxide film. However, the much faster attack found in wet corrosion is due to the following:

(a) The $Fe(OH)_2$ either deposits *away* from the corroding material; or, if it deposits on the surface, it does so as a loose deposit, giving little or no protection.
(b) Consequently M^{++} and OH^- usually diffuse in the *liquid* state, and therefore do so very rapidly.
(c) In *conducting* materials, the electrons can move very easily as well.

The result is that the oxidation of iron in aerated water (rusting) goes on at a rate which is millions of times faster than that in dry air. Because of the importance of (c), wet oxidation is a particular problem with metals.

Voltage differences as a driving force for wet oxidation

In dry oxidation we quantified the tendency for a material to oxidise in terms of the energy needed, in kJ mol^{-1} of O_2, to manufacture the oxide from the material and oxygen. Because wet oxidation involves electron flow in conductors, which is easier to

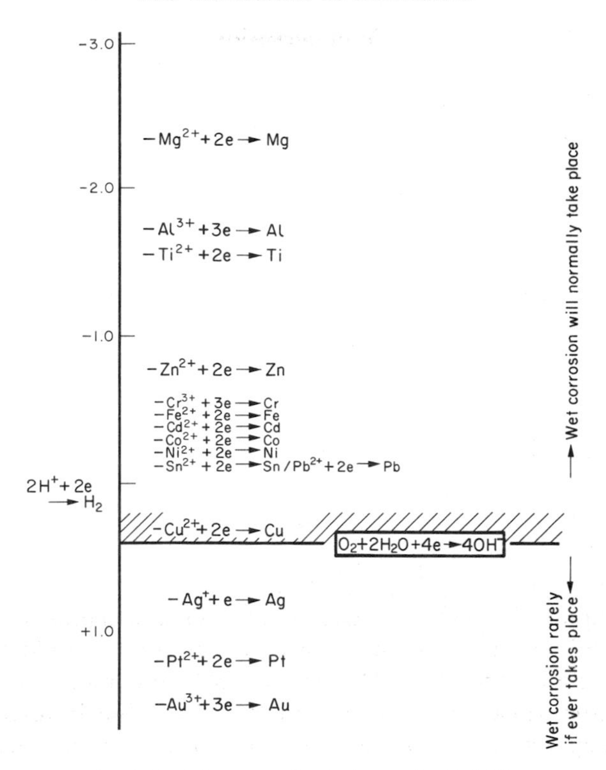

Fig. 23.3. Wet corrosion voltages (at 300 K).

measure, the tendency of a metal to oxidise in solution is described by using a *voltage* scale rather than an *energy* one.

Figure 23.3 shows the voltage differences that would just stop various metals oxidising in aerated water. As we should expect, the information in the figure is similar to that in our previous bar-chart (see Chapter 21) for the *energies* of oxidation. There are some differences in ranking, however, due to the differences between the detailed reactions that go on in dry and wet oxidation.

What do these voltages mean? Suppose we could separate the cathodic and the anodic regions of a piece of iron, as shown in Fig. 23.4. Then at the cathode, oxygen is reduced to OH^-, absorbing electrons, and the metal therefore becomes positively charged. The reaction continues until the potential rises to +0.401 V. Then the coulombic attraction between the +ve charged metal and the −ve charged OH^- ion becomes so large that the OH^- is pulled back to the surface, and reconverted to H_2O and O_2; in other words, the reaction stops. At the anode, Fe^{++} forms, leaving electrons behind in the metal which acquires a negative charge. When its potential falls to −0.440 V, that reaction, too, stops (for the same reason as before). If the anode and cathode are now *connected*, electrons flow from the one to the other, the potentials fall, and both reactions start up again. The difference in voltage of 0.841 V is the driving potential for the oxidation reaction. The bigger it is, the bigger the tendency to oxidise.

Now a note of caution about how to interpret the voltages. For convenience, the voltages given in reference books always relate to ions having certain specific concentrations (called "unit activity" concentrations). These concentrations are high—and

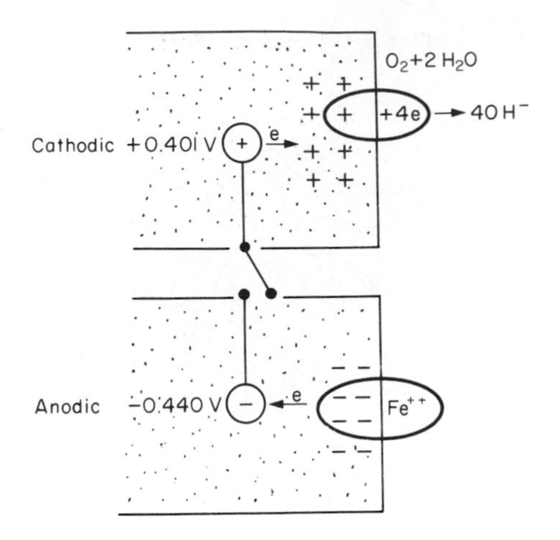

Fig. 23.4. The voltages that drive wet corrosion.

make it rather hard for the metals to dissolve (Fig. 23.5). In dilute solutions, metals can corrode more easily, and this sort of effect tends to move the voltage values around by up to 0.1 V or more for some metals. The important thing about the voltage figures given therefore is that they are only a *guide* to the driving forces for wet oxidation.

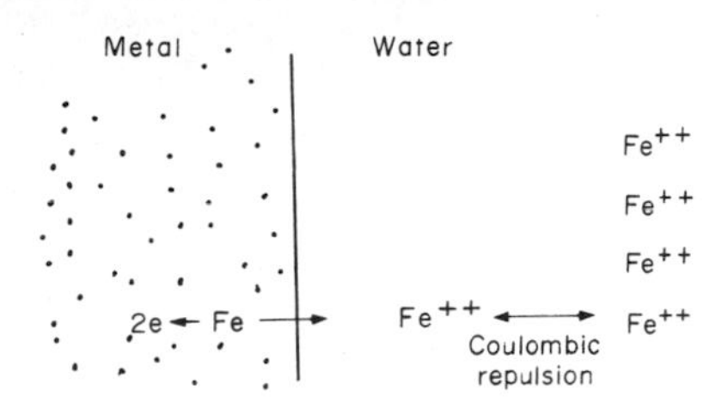

Fig. 23.5. Corrosion takes place less easily in concentrated solution.

Obviously, it is not very easy to measure voltage variations *inside* a piece of iron, but we can artificially transport the "oxygen-reduction reaction" away from the metal by using a piece of metal that does not normally undergo wet oxidation (e.g. platinum) and which serves merely as a *cathode* for the oxygen-reduction reaction.

The corrosion voltages of Fig. 23.3 also tell you what will happen when two *dissimilar* metals are joined together and immersed in water. If copper is joined to zinc, for instance, the zinc has a larger corrosion voltage than the copper. The zinc therefore becomes the anode, and is attacked; the copper becomes the cathode, where the oxygen reaction takes place, and it is unattacked. Such *couples* of dissimilar metals can be dangerous: the attack at the anode is sometimes very rapid, as we shall see in the next chapter.

Rates of wet oxidation

As one might expect on the basis of what we said in the chapters on dry oxidation, the rates of wet oxidation found in practice bear little relationship to the voltage driving forces for wet oxidation, provided these are such that the metal is prone to corrosion in the first place. To take some examples, the approximate surface losses of some metals in mm per year in clean water are shown on Fig. 23.6. They are almost the reverse of

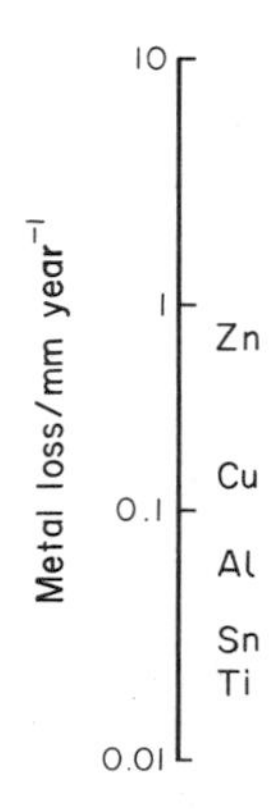

Fig. 23.6. Corrosion rates of some metals in clean water.

the order expected in terms of the voltage driving forces for wet oxidation. The slow rate of wet oxidation for Al, for example, arises because it is very difficult to prevent a thin, dry oxidation film of Al_2O_3 forming on the metal surface. In *sea* water, on the other hand, Al corrodes very rapidly because the chloride ions tend to break down the protective Al_2O_3 film. Because of the effect of "foreign" ions like this in most practical environments, corrosion rates vary very widely indeed for most materials. Materials Handbooks often list rough figures of the wet oxidation resistance of metals and alloys in various environments (ranging from beer to sewage!).

Localised attack: corrosion cracking

It is often found that wet corrosion attacks metals *selectively* as well as, or instead of, uniformly, and this can lead to component failure *much* more rapidly and insidiously than one might infer from average corrosion rates (Fig. 23.7). Stress and corrosion

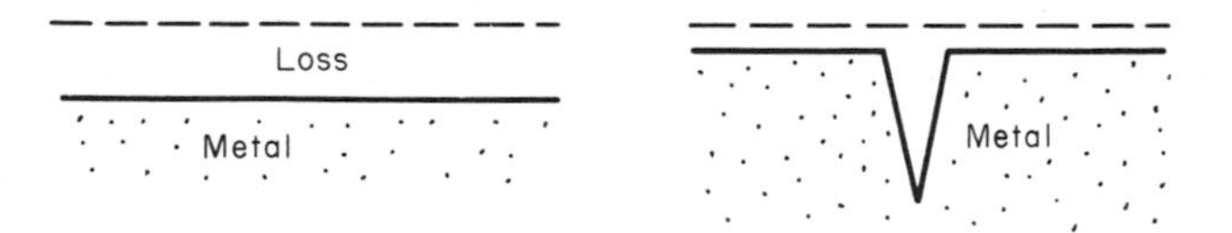

Fig. 23.7. Localised attack.

acting together can be particularly bad, giving cracks which propagate rapidly and unexpectedly. Four types of corrosion cracking commonly lead to unplanned failures. These are:

(a) *Stress corrosion cracking*

In some materials and environments, cracks grow steadily under a constant stress intensity K which is much less than K_c (Fig. 23.8). This is obviously dangerous: a

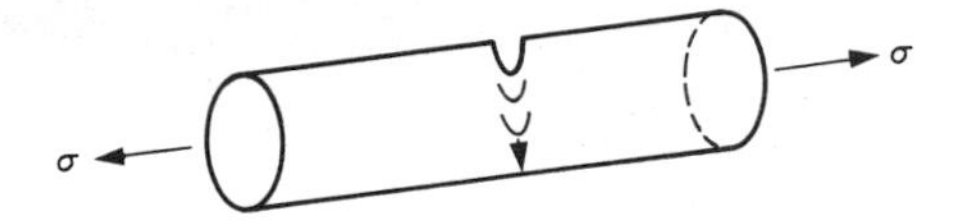

Fig. 23.8. Stress corrosion cracking.

structure which is safe when built can become unsafe with time. Examples are brass in ammonia, mild steel in caustic soda, and some Al and Ti alloys in salt water.

(b) *Corrosion fatigue*

Corrosion increases the rate of growth of fatigue cracks in most metals and alloys, e.g. the stress to give $N_f = 5 \times 10^7$ cycles decreases by 4 times in salt water for many steels (Fig. 23.9). The crack growth rate is larger—often much larger—than the sum of the rates of corrosion and fatigue, each acting alone.

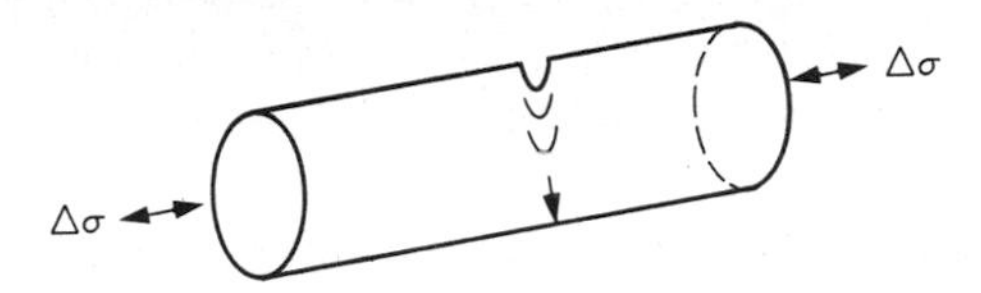

Fig. 23.9. Corrosion fatigue.

(c) *Intergranular attack*

Grain boundaries have different corrosion properties from the grains and may corrode preferentially, giving cracks that then propagate by stress corrosion or corrosion fatigue (Fig. 23.10).

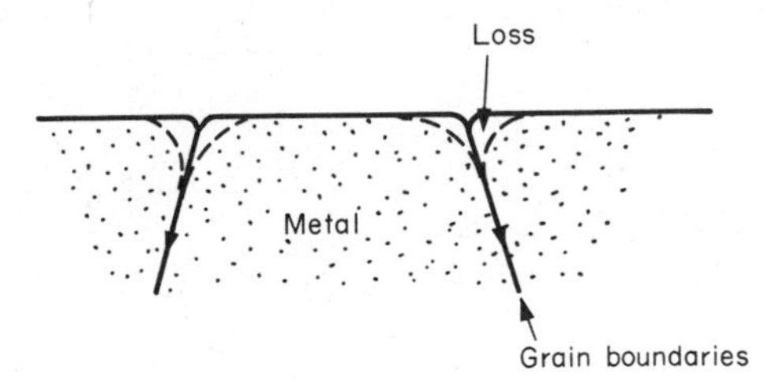

Fig. 23.10. Intergranular attack.

(d) *Pitting*

Preferential attack can also occur at breaks in the oxide film (caused by abrasion), or at precipitated compounds in certain alloys (Fig. 23.11).

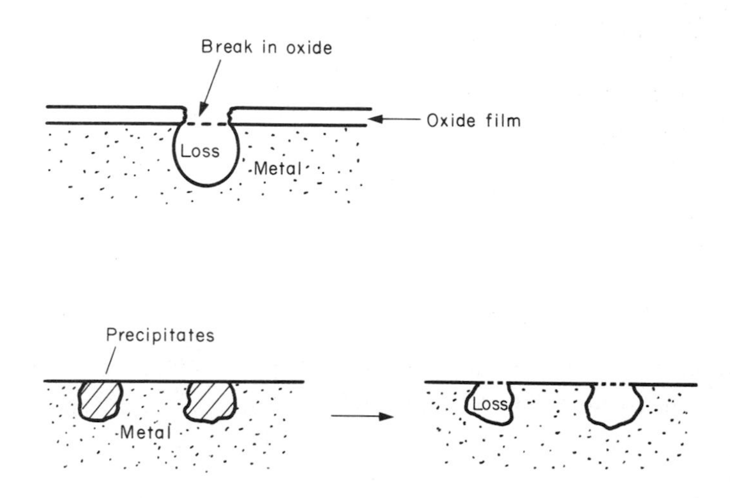

Fig. 23.11. Pitting corrosion.

To summarise, unexpected corrosion failures are much more likely to occur by localised attack than by uniform attack (which can easily be detected); and although corrosion handbooks are useful for making initial choices of materials for applications where corrosion is important, critical components must be checked for life-to-fracture in closely controlled experiments resembling the actual environment as nearly as possible.

In the next chapter we shall look at some case studies in corrosion-resistant designs which are based on the ideas we have just discussed.

Further reading

J. P. Chilton, *Principles of Metallic Corrosion*, 2nd edition, The Chemical Society, 1973, Chap. 3.
M. G. Fontana and N. D. Greene, *Corrosion Engineering*, McGraw Hill, 1967, Chaps. 2 and 3.
J. C. Scully, *The Fundamentals of Corrosion*, 2nd edition, Pergamon Press, 1975, Chap. 2.
C. J. Smithells, *Metals Reference Book*, 5th edition, Butterworths, 1976 (for data).
Metals Handbook, 8th edition, American Society for Metals, 1975, Vol. 10 (for data).

CHAPTER 24

CASE STUDIES IN WET CORROSION

Introduction

We now examine three real corrosion problems: the protection of pipelines, the selection of a material for a factory roof, and materials for car exhaust systems. The rusting of iron appears in all three case studies, but the best way of overcoming it differs in each. Sometimes the best thing is to change to a new material which does not rust; but often economics prevent this, and ways must be found to slow down or stop the rusting reaction.

Case study 1: the protection of underground pipes

Many thousands of miles of steel pipeline have been laid under, or in contact with, the ground for the long-distance transport of oil, natural gas, etc. Obviously corrosion is a problem if the ground is at all damp, as it usually will be, and if the depth of soil is not so great that oxygen is effectively excluded. Then the oxygen reduction reaction

$$O_2 + 2H_2O + 4e \rightarrow 4OH^-$$

and the metal-corroding reaction

$$Fe \rightarrow Fe^{++} + 2e$$

can take place, causing the pipe to corrode. Because of the capital cost of pipelines, their inaccessibility if buried, the disruption to supplies caused by renewal, and the potentially catastrophic consequences of undetected corrosion failure, it is obviously very important to make sure that pipelines do not corrode. How is this done?

One obvious way of protecting the pipe is by covering it with some inert material to keep water and oxygen out: thick polyethylene sheet stuck in position with a butyl glue, for example. The end sections of the pipes are left uncovered ready for welding—and the welds are subsequently covered on site. However, such coverings rarely provide complete protection—rough handling on site frequently leads to breakages of the film, and careless wrapping of welds leaves metal exposed. What can we do to prevent localised attack at such points?

Sacrificial protection

If the pipe is connected to a slab of material which has a more negative corrosion voltage (Fig. 24.1), then the couple forms an electrolytic cell. As explained in Chapter 23, the more electronegative material becomes the anode (and dissolves), and the pipe becomes the cathode (and is protected).

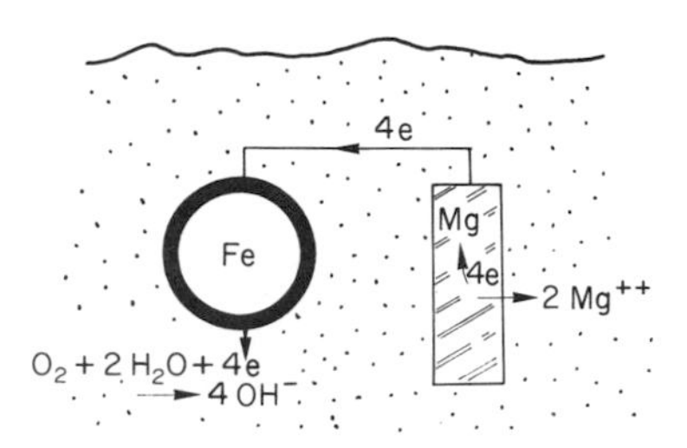

Fig. 24.1. Sacrificial protection of pipelines. Typical materials used are Mg (with 6% Al, 3% Zn, 0.2% Mn), Al (with 5% Zn) and Zn.

As Fig. 24.1 shows, pipelines are protected from corrosion by being wired to anodes in just this way. Magnesium alloy is often used because its corrosion voltage is very low (much lower than that of zinc) and this attracts Fe^{++} to the steel very strongly; but aluminium alloys and zinc are used widely too. The alloying additions help prevent the formation of a protective oxide on the anode—which might make it become cathodic. With some metals in particular environments (e.g. titanium in sea water) the nature of the oxide film is such that it effectively prevents metal passing through the film into solution. Then, although titanium is very negative with respect to iron (see Fig. 23.3), it fails to protect the pipeline (Fig. 24.2). Complications like this can also affect other metals (e.g. Al, Cd, Zn) although generally to a much smaller extent. *This sort of behaviour is another reason for our earlier warning that corrosion voltages are only general guides to corrosion behaviour*—again, experimental work is usually a necessary prelude to design against corrosion.

Naturally, because the protection depends on the dissolution of the anodes, these require replacement from time to time (hence the term "sacrificial" anodes). In order to minimise the loss of anode metal, it is important to have as good a barrier layer around the pipe as possible, even though the pipe would still be protected with no barrier layer at all.

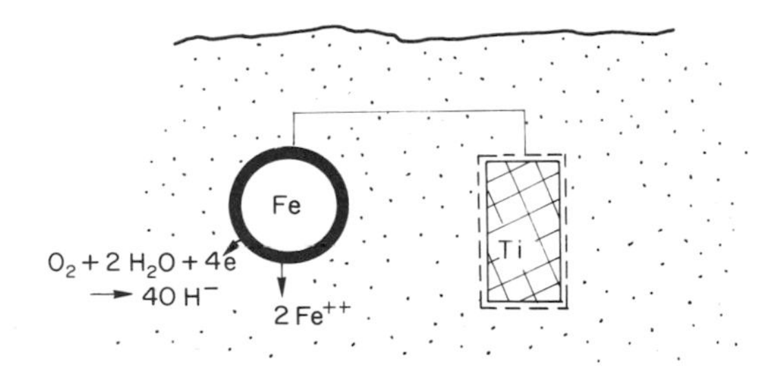

Fig. 24.2. Some sacrificial materials do not work because they carry a "passivating" oxide layer.

Protection by imposing a potential

An alternative way of protecting the pipe is shown on Fig. 24.3. Scrap steel is buried near the pipe and connected to it through a battery or d.c. power supply, which maintains a sufficient potential difference between them to make sure that the scrap is

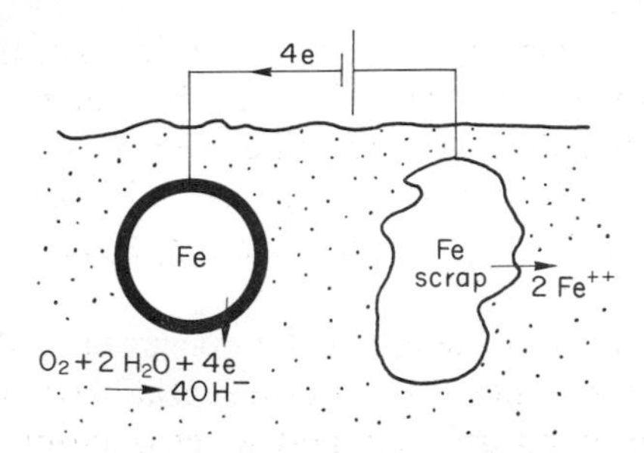

Fig. 24.3. Protection of pipelines by imposed potential.

always the anode and the pipe the cathode (it takes roughly the corrosion potential of iron—a little under 1 volt). This alone will protect the pipe, but unless the pipe is coated, a large current will be needed to maintain this potential difference.

Alternative materials

Cost rules out almost all alternative materials for long-distance pipe lines: it is much cheaper to build and protect a mild steel pipe than to use stainless steel instead—even though no protection is then needed. The only competing material is a polymer, which is completely immune to wet corrosion of this kind. City gas mains are now being replaced by polymeric ones; but for large diameter transmission lines, the mechanical strength of steel makes it the preferred choice.

CASE STUDY 2: MATERIALS FOR A LIGHTWEIGHT FACTORY ROOF

Let us now look at the corrosion problems that are involved in selecting a material for the lightweight roof of a small factory. Nine out of ten people asked to make a selection would think first of corrugated, *galvanised steel*. This is strong, light, cheap and easy to install. Where's the catch? Well, fairly new galvanised steel is rust-free, but after 20 to 30 years, rusting sets in and the roof eventually fails.

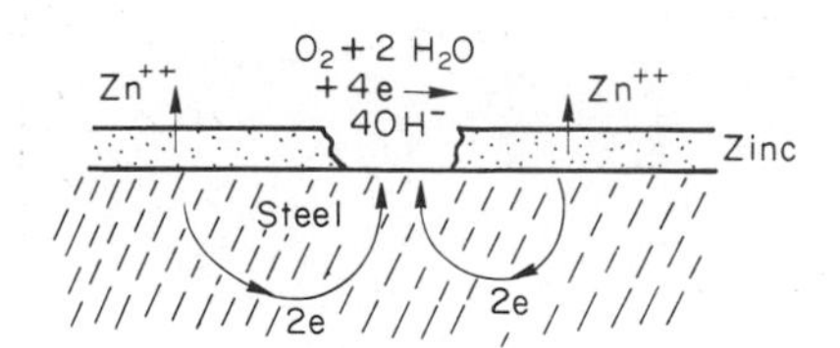

Fig. 24.4. Galvanised steel is protected by a sacrificial layer of zinc.

How does galvanising work? As Fig. 24.4 shows, the galvanising process leaves a thin layer of zinc on the surface of the steel. This acts as a barrier between the steel and the atmosphere; and although the driving voltage for the corrosion of zinc is greater than that for steel (see Fig. 23.3) in fact zinc corrodes quite slowly in a normal urban atmosphere because of the barrier effect of its oxide film. The loss in thickness is typically 0.1 mm in 20 years.

If scratches and breaks occur in the zinc layers by accidental damage—which is certain to occur when the sheets are erected—then the zinc will cathodically protect the iron (see Fig. 24.4) in exactly the way that pipelines are protected using zinc anodes. This explains the long postponement of rusting. But the coating is only about 0.15 mm thick, so after about 30 years most of the zinc has gone, rusting suddenly becomes chronic, and the roof fails.

At first sight, the answer would seem to be to increase the thickness of the zinc layer. This is not easily done, however, because the hot dipping process used for galvanising is not sufficiently adjustable; and electroplating the zinc onto the steel sheet increases the production cost considerably. Painting the sheet (for example, with a bituminous paint) helps to reduce the loss of zinc considerably, but at the same time should vastly decrease the area available for the cathodic protection of the steel; and if a scratch penetrates both the paint and the zinc, the exposed steel may corrode through much more quickly than before.

Alternative materials

A relatively recent innovation has been the architectural use of *anodised aluminium*. Although the driving force for the wet oxidation of aluminium is very large, aluminium corrodes very slowly in fresh-water environments because it carries a very adherent film of the poorly conducting Al_2O_3. In anodised aluminium, the Al_2O_3 film is artificially thickened in order to make this barrier to corrosion extremely effective. In the anodising process, the aluminium part is put into water containing various additives to promote compact film growth (e.g. boric acid). It is then made positive electrically which attracts the oxygen atoms in the polar water molecules (see Chapter 4). The attached oxygen atoms react continuously with the metal to give a thickened oxide film as shown on Fig. 24.5. The film can be coloured for aesthetic purposes by adding colouring agents towards the end of the process and changing the composition of the bath to allow the colouring agents to be incorporated.

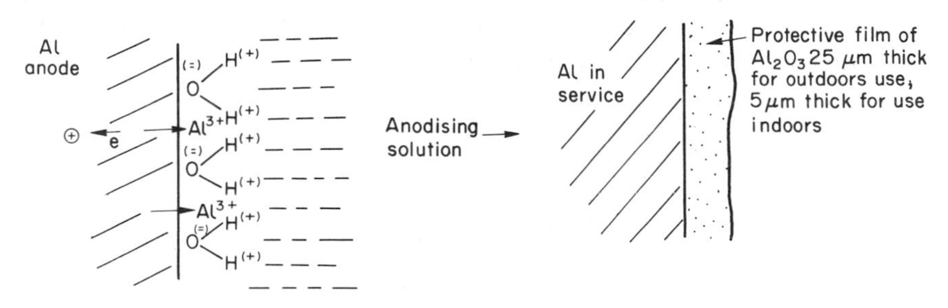

Fig. 24.5. Protecting aluminium by anodising it.

Finally, what of polymeric materials? Corrugated plastic sheet is commonly used for roofing small sheds, car ports and similar buildings; but although polymers do not generally corrode—they are often used in nasty environments like chemical plant—they *are* prone to damage by the ultraviolet wavelengths of the sun's radiation. These high-energy photons, acting over a period of time, gradually break up the molecular chains in the polymer and degrade its mechanical properties.

A note of caution about roof fasteners. A common mistake is to fix a galvanised or aluminium roof in place with nails or screws of a *different* metal: copper or brass, for instance. The copper acts as cathode, and the zinc or aluminium corrodes away rapidly near to the fastening. A similar sort of goof has been known to occur when copper roofing sheet has been secured with steel nails. As Fig. 24.6 shows, this sort of situation leads to catastrophically rapid corrosion not only because the iron is anodic, but because it is so easy for the electrons generated by the anodic corrosion to get away to the large copper cathode.

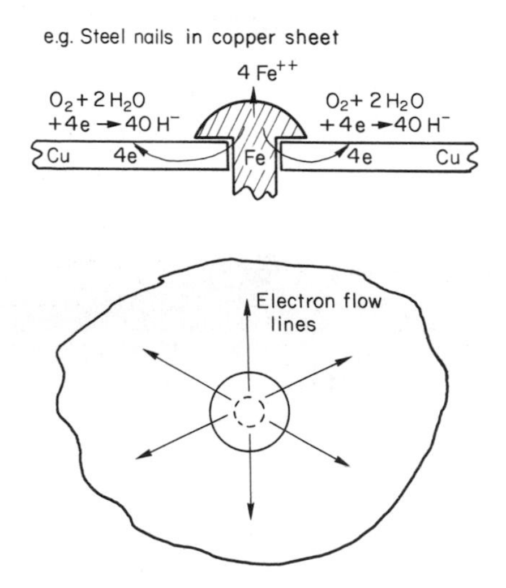

Fig. 24.6. Large cathodes can lead to very rapid corrosion.

Case study 3: automobile exhaust systems

The lifetime of a conventional exhaust system on an average family car is only 2 years or so. This is hardly surprising—mild steel is the usual material and, as we have shown, it is not noted for its corrosion resistance. The interior of the system is not painted and begins to corrode immediately in the damp exhaust gases from the engine. The single coat of cheap cosmetic paint soon falls off the outside and rusting starts there, too, aided by the chloride ions from road salt, which help break down the iron oxide film.

The lifetime of the exhaust system could be improved by galvanising the steel to begin with. But there are problems in using platings where steel has to be joined by *welding*. Zinc, for example, melts at 420°C and would be burnt off the welds; and breaks would still occur if plating metals of higher melting point (e.g. Ni, 1455°C) were used. Occasionally manufacturers fit chromium-plated exhaust systems but this is for appearance only: if the plating is done before welding, the welds are unprotected and will corrode quickly; and if it is done after welding, the interior of the system is unplated and will corrode.

Alternative materials

The most successful way of combating exhaust-system corrosion is, in fact, stainless steel. This is a good example of how—just as with dry oxidation—the addition of foreign atoms to a metal can produce stable oxide films that act as barriers to corrosion. In the case of stainless steel, Cr is dissolved in the steel in solid solution, and Cr_2O_3 forms on the surface of the steel to act as a corrosion barrier.

There is one major pitfall which must be avoided in using stainless-steel components joined by welding: it is known as *weld decay.* It is sometimes found that the *heat-affected zone*—the metal next to the weld which got hot but did not melt—corrodes badly.

Figure 24.7 explains why. All steels contain carbon—for their mechanical properties—and this carbon can "soak up" chromium (at grain boundaries in particular) to form precipitates of the compound *chromium carbide.* Because the regions near

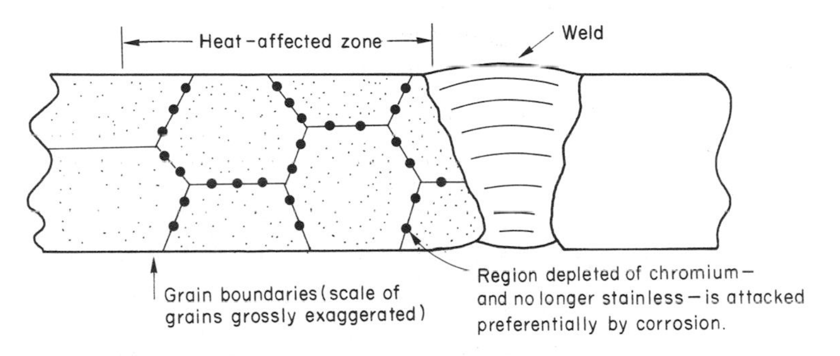

Fig. 24.7. Weld decay in stainless steel.

the grain boundaries lose most of their chromium in this way, they are no longer protected by Cr_2O_3, and corrode badly. The cure is to *stabilise* the stainless steel by adding Ti or Nb which soaks up the carbon near the grain boundaries.

Further reading

M. G. Fontana and N. D. Greene, *Corrosion Engineering*, McGraw Hill, 1967.
D. R. Gabe, *Principles of Metal Surface Treatment and Protection*, 2nd edition, Pergamon Press, 1978.
R. D. Barer and B. F. Peters, *Why Metals Fail*, Gordon & Breach, 1970.
Metals Handbook, 8th edition, American Society for Metals, 1975, Vol. 10.

G. Friction, abrasion and wear

CHAPTER 25

FRICTION AND WEAR

Introduction

We now come to the final properties that we shall be looking at in this book on engineering materials: the frictional properties of materials in contact, and the wear that results from such contact. This is of considerable importance in most of mechanical engineering design. In the case of bearing surfaces, frictional forces are undesirable because of the power they waste; and wear is bad because it leads to poor working tolerances. On the other hand, when selecting materials for clutch and brake linings—or even for the soles of our shoes—we aim to achieve maximum friction; but again, wear is undesirable for obvious reasons. Finally, in metal-working operations such as milling and grinding we try to achieve maximum wear with the minimum of energy expended in friction. In Chapters 25 and 26 we shall examine the origins of friction and wear and then go on to look at some case studies which illustrate the influence of friction and wear on component design.

Friction between materials

As you know, when two materials are placed in contact, any attempt to cause one of the materials to slide over the other is resisted by a frictional force (Fig. 25.1). The

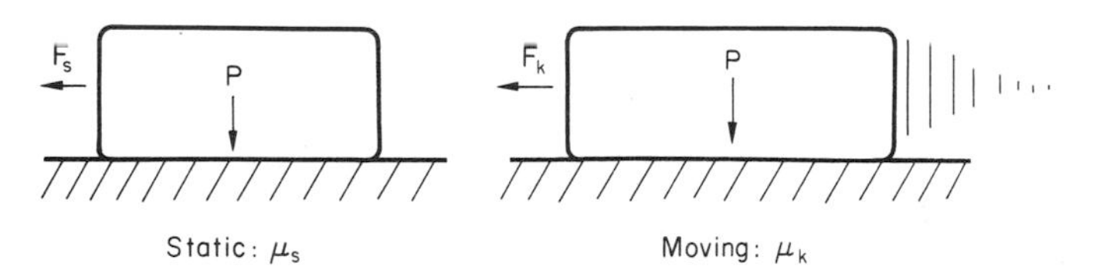

Fig. 25.1. Static and kinetic coefficients of friction.

force that will just cause sliding to start, F_s, is related to the normal force P across the contact surface by

$$F_s = \mu_s P \tag{25.1}$$

where μ_s is the coefficient of static friction.

Once sliding starts, the limiting frictional force decreases slightly and we can write

$$F_k = \mu_k P \tag{25.2}$$

where μ_k ($<\mu_s$) is the coefficient of kinetic friction (Fig. 25.1).

These results at first sight run counter to our intuition—how is it that the friction between contacting materials can depend only on the *force* pressing them together—and not apparently on the area? In order to understand this behaviour in metals, we must first look at the geometry of a typical metal surface.

If the surface of a fine-turned bar of copper is examined by making an oblique slice through it (a "taper section" which magnifies the height of any asperities), or if its profile is measured with a "Talysurf" (a device like a gramophone pick-up which, when run across a surface, plots out the hills and valleys), it is found that the surface looks like Fig. 25.2.

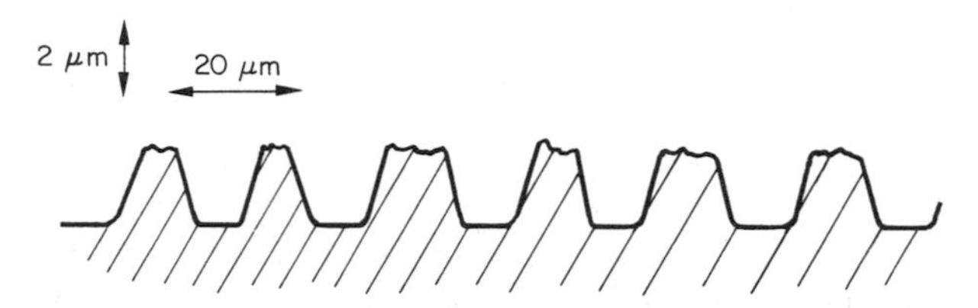

Fig. 25.2. What a finely machined metal surface looks like at high magnification (the heights of the asperities are plotted on a much more exaggerated scale than the lateral distances between asperities).

You can clearly see that the surface consists of a large number of projections or *asperities*—it looks rather like a cross-section through Switzerland. If the metal is abraded with the finest abrasive paper, the scale of the asperities decreases by ten times but the asperities are still well formed. Even if the metal is polished for a long time using the finest type of metal polish, asperities still survive.

When two nominally flat surfaces are placed in contact, no matter how carefully they have been machined and polished, they will only contact where one set of asperities touches the other—it is rather like turning Austria upside down and putting it on Switzerland. Any load across the surfaces is supported solely by the places where the asperities make contact, and therefore only a very small fraction of the *nominal* area of the surfaces carries the load.

Initially, at very low load, the asperities deform elastically where they touch. However, for realistic loads, extensive *plastic* deformation takes place at the tips of asperities. If each asperity yields, forming a junction across the surface, then the total load transmitted across the surface (Fig. 25.3) is

$$P \approx a\sigma_y, \tag{25.3}$$

where a is the *actual* area of contact, and σ_y is the compressive yield stress. In other words, the real area of contact is given by

$$a \approx \frac{P}{\sigma_y}. \tag{25.4}$$

Obviously, if we double P we double the real area of contact, a.

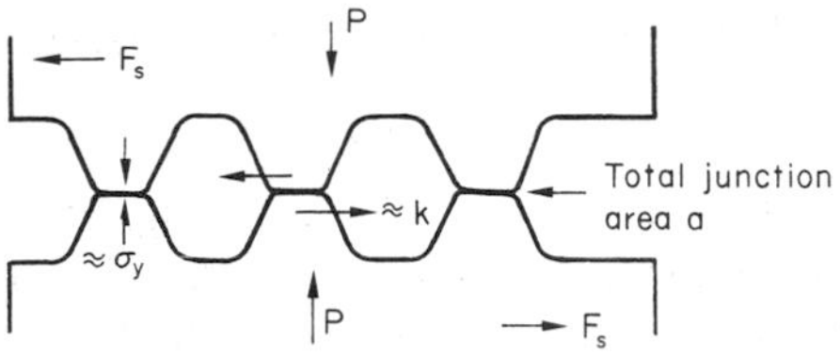

Fig. 25.3.

Let us now look at how this contact geometry influences the friction between metal surfaces in contact. If you attempt to slide one of the surfaces over the other, the sliding is opposed by a shear stress in the asperities. This shear stress, τ, is greatest where the cross-sectional area of asperities is least, i.e. is greatest in the region of contact. The force resisting sliding is given by

$$F = a\tau.$$

Now, the intense plastic deformation in the regions of contact presses the asperity tips together so well that there is atom-to-atom contact across the area of contact a. The junction, therefore, can withstand a shear stress as large as k approximately, where k is the shear-yield stress of the material (Chapter 11). The force, F_s, at which sliding takes place is therefore

$$F_s \approx ak \approx a\sigma_y/2. \tag{25.5}$$

Combining this result with our first equation, we have

$$F_s \approx \frac{P}{2}. \tag{25.6}$$

This is simply a statement of the law of friction

$$F_s = \mu_s P$$

which we have thus deduced from first principles using our simple model of asperity contact. Our model also predicts that $\mu_s \approx \frac{1}{2}$, which is of the right order for the coefficients of static friction between metal surfaces.

How do we explain the lower value of μ_k? Well, once the surfaces are sliding, there is not as much *time* available for atom-to-atom contact to occur at the asperity junctions as when the surfaces are in static contact, and the contact area over which shearing needs to take place is correspondingly reduced. As soon as sliding stops, creep allows the contacts to grow a little, and diffusion allows the bond there to become stronger, and μ rises again to μ_s.

Data for coefficients of friction

If *metal* surfaces are thoroughly cleaned in vacuum it is almost impossible to slide them over each other. Any shearing force causes further plasticity at the junctions, which quickly grow, leading to complete seizure ($\mu > 5$, Table 25.1). This is a problem

TABLE 25.1
COEFFICIENTS OF FRICTION

Material	μ
Perfectly clean metals in vacuum	Seizure $\mu > 5$
Clean metals in air	0.8–2
Clean metals in wet air	0.5–1.5
Steel on dry bearing metals (e.g. lead, bronze)	0.1–0.5
Steel on ceramics (e.g. sapphire, diamond, ice)	0.1–0.5
Ceramics on ceramics (e.g. carbides on carbides)	0.05–0.5
Polymers on polymers	0.05–1.0
Metals and ceramics on polymers (PE, PTFE, PVC)	0.04–0.5
Boundary lubrication of metals	0.05–0.2
High-temperature lubricants (MoS_2, graphite)	0.05–0.2
Hydrodynamic lubrication	0.001–0.005

in outer space, and in atmospheres (e.g. H_2) which remove any surface films from the metal. The smallest trace of oxygen or H_2O greatly reduces μ by creating an oxide film which prevents these extremely large metallic junctions forming.

We said in Chapter 21 that all metals except gold have a layer, no matter how thin, of metal oxide on their surfaces. Experimentally, it is found that for some metals the junction between the oxide films formed at asperity tips is weaker in shear than the neighbouring metal (Fig. 25.4). In this case, sliding of the surfaces will take place in the thin oxide layer, at a stress less than in the metal itself, and lead to a corresponding reduction in μ to between 0.5 and 1.5.

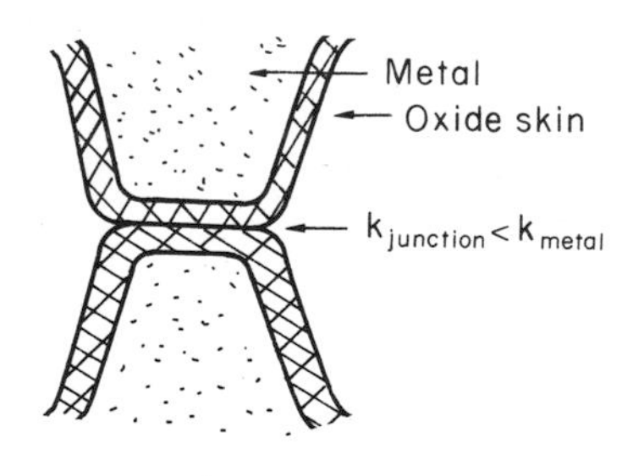

Fig. 25.4.

When soft metals slide over each other (e.g. lead on lead) the junctions are weak but their area is large (eqn. (25.4))—so μ is large (0.5 to 1.5). When hard metals slide (e.g. steel on steel) the junctions are small, but they are strong, and again friction is large. Many bearings are made of a thin film of a soft metal between two hard ones, giving weak junctions of small area (see Chapter 26). *White metal* bearings, for example, consist of soft alloys of lead or tin supported in a matrix of stronger phases; *bearing bronzes* consist of soft lead particles (which smear out to form the lubricating film) supported in a bronze matrix; and polymer-impregnated *porous bearings* are made by partly sintering copper with a polymer (usually PTFE) forced into its pores.

Bearings like these are not designed to run dry—but if lubrication does break down, the soft component gives a coefficient of friction of 0.1 to 0.5 which may be low enough to prevent catastrophic overheating and seizure.

When metals slide on bulk polymers, friction is still caused by adhesive junctions. But any plastic flow tends to orient the polymer chains parallel to the sliding surface, and in this orientation they shear easily and μ is low—0.05 to 0.2. Polymers make attractive low-friction bearings, although they have some drawbacks: polymer molecules peel easily off the sliding surface, so wear is heavy; and because creep allows junction growth when the slider is stationary, the coefficient of static friction, μ_s, is sometimes much larger than that for sliding friction, μ_k.

Lubrication

As we said in the introduction, friction absorbs a lot of work in machinery and—as well as wasting power—this work is mainly converted to *heat* at the sliding surfaces, which can damage and even melt the bearing. In order to minimize frictional forces we need to make it as easy as possible for surfaces to slide over one another, and the obvious way to try to do this is to contaminate the asperity tips with something that: (a) can stand the pressure at the bearing surface and so prevent atom-to-atom contact between asperities; (b) can itself shear easily.

Usually, however, we would like a much larger reduction in μ than that given by soft films or polymers, and then we must use *lubricants*. The standard lubricants are oils, greases and fatty materials such as soap and animal fats. These "contaminate" the surfaces, preventing adhesive contact; and the thin layer of oil or grease shears easily when a shearing force F_s is applied, and so obviously lowers the coefficient of friction. What is *not* so obvious is why the very fluid oil is not squeezed out from between the asperities by the enormous pressures generated there. Well, oils nowadays have small amounts ($\approx 1\%$) of active organic molecules added. One end of each molecule reacts with the metal oxide surface and sticks to it, whereas the other ends attract one another to form an oriented "forest" of molecules (Fig. 25.5). These forests can resist very large forces normal to the surface, and hence separate the asperity tips very effectively; whilst the two layers of molecules can shear over themselves quite easily. This type of lubrication is termed *boundary lubrication*, and is capable of reducing μ by a factor of

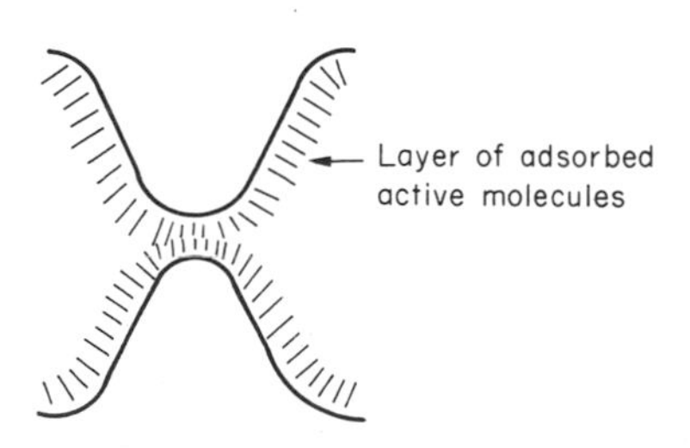

Fig. 25.5. Boundary lubrication.

10 (Table 25.1). *Hydrodynamic lubrication* is even more effective: we shall discuss it in the next chapter.

Even the best boundary lubricants cease to work above about 200°C. Soft metal bearings like those described above can cope with *local* hot spots: the soft metal melts and provides a local lubricating film. But when the entire bearing runs hot, special lubricants are needed. The best are a suspension of PTFE in oil (good to 320°C); graphite (good to 600°C); and molybdenum disulphide (good to 800°C).

Wear of materials

Even when solid surfaces are protected by oxide films and boundary lubricants, some solid-to-solid contact occurs at regions where the oxide film breaks down under mechanical loading, and adsorption of active boundary lubricants is poor. This intimate contact will generally lead to *wear*. Wear is normally divided into two main types: *adhesive wear* and *abrasive wear*.

Adhesive wear

Figure 25.6 shows that, if the adhesion between A atoms and B atoms is good enough, wear fragments will be removed from the softer metal A. If materials A and B

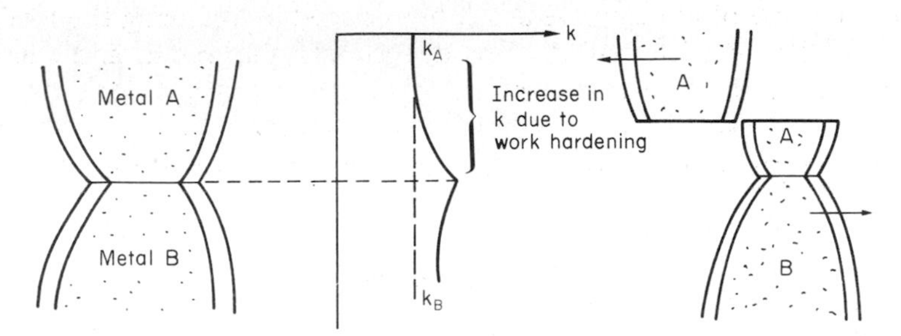

Fig. 25.6. Adhesive wear.

are the same, wear takes place from *both* surfaces. The size of the piece of metal removed from a particular asperity depends on how far away from the junction the shearing takes place. If the work-hardened region extends well into the asperity, the tendency will be to produce large pieces; on the other hand, this tendency will be reduced if the cross-section of the asperity increases away from the contact region.

In order to minimise the *rate of wear* we obviously need to minimise the size of each piece of metal removed. The obvious way to do this is to minimise the area of contact a. Since $a \approx P/\sigma_y$, reducing the loading on the surfaces will reduce the wear, as would seem intuitively obvious. You can demonstrate this by writing with chalk on the blackboard. At high load there is plenty of transfer of chalk to the board, indicating severe wear. At low load, little wear takes place. (So, you see, wear is sometimes useful!) The second way to reduce a is to increase σ_y, i.e. the *hardness*. This is why hard pencils write less clearly than soft pencils.

Abrasive wear

Wear fragments produced by adhesive wear often become detached from their asperities during further sliding of the surfaces. Because oxygen is desirable in lubricants (to help maintain the oxide-film barrier between the sliding metals) these detached wear fragments can become oxidised to give hard oxide particles which *abrade* the surfaces in the way that sandpaper might.

Figure 25.7 shows how a hard material can "plough" wear fragments from a softer material, producing severe abrasive wear. Abrasive wear is not, of course, confined to indigenous wear fragments, but can be caused by dirt particles (e.g. sand) making their way into the system, or—in an engine—by combustion products: that is why it is important to filter the oil.

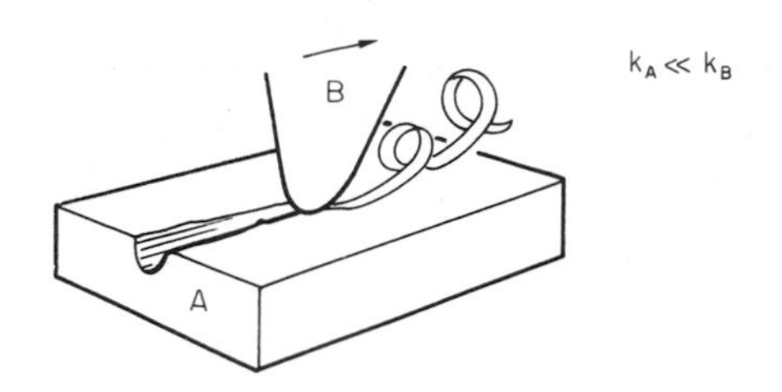

Fig. 25.7. Abrasive wear.

Obviously, the rate of abrasive wear can be reduced by reducing the load—just as in a hardness test. The particle will dig less deeply into the metal, and cause a smaller furrow to be ploughed. Increasing the hardness of the metal will have the same effect. Again, although abrasive wear is usually bad—as in machinery—we would find it difficult to sharpen lathe tools, or polish brass ornaments, or drill rock, without it.

Further reading

F. P. Bowden and D. Tabor, *Friction—An Introduction to Tribology*, Heinemann Science Study Series, No. 41, 1974.

F. P. Bowden and D. Tabor, *The Friction and Lubrication of Solids*, Oxford University Press, Part I, 1950; Part II, 1965.

A. D. Sarkar, *Wear of Metals*, Pergamon Press, 1976.

E. Rabinowicz, *Friction and Wear of Materials*, Wiley, 1965.

C. J. Smithells, *Metals Reference Book*, 5th edition, Butterworths, 1976.

CHAPTER 26

CASE STUDIES IN FRICTION AND WEAR

Introduction

In this chapter we examine three quite different problems involving friction and wear. The first involves most of the factors that appeared in Chapter 25: it is that of a round shaft or journal rotating in a cylindrical bearing. This type of *journal bearing* is common in all types of rotating or reciprocating machinery: the crankshaft bearings of an automobile are good examples. The second is quite different, and introduces the frictional properties of ice in the design of skis and sledge runners. The third case study introduces us to some of the frictional properties of polymers: the selection of rubbers for anti-skid tyres.

Case study 1: the design of journal bearings

In the proper functioning of a well-lubricated journal bearing, the frictional and wear properties of the materials are, surprisingly, irrelevant. This is because the mating surfaces are kept apart by a thin pressurised film of oil formed under conditions of *hydrodynamic lubrication*. Figure 26.1 shows a cross-section of a bearing operating hydrodynamically. The load on the journal pushes the shaft to one side of the bearing,

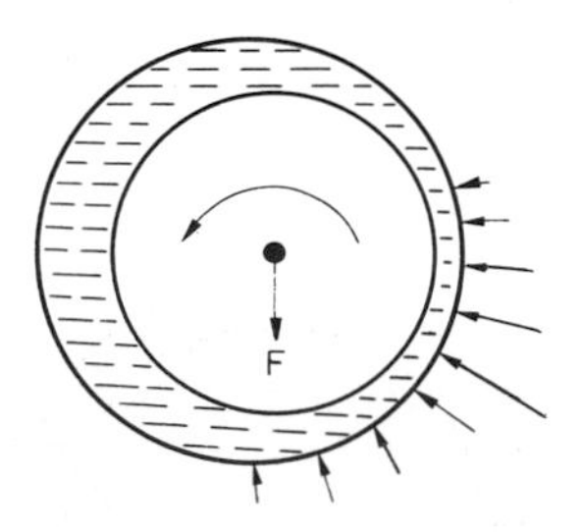

Fig. 26.1. Hydrodynamic lubrication.

so that the working clearance is almost all concentrated on one side. Because oil is viscous, the revolving shaft drags oil around with it. The convergence of the oil stream towards the region of nearest approach of the mating surfaces causes an increase in the pressure of the oil film, and this pressure literally holds up the shaft against the applied force. Pressures of 10 to 100 atmospheres are common under such conditions. Provided the oil is sufficiently viscous, the film at its thinnest region is still thick enough to cause complete separation of the mating surfaces: there should be no asperity contact, and no wear, under ideal hydrodynamic conditions. In addition to this absence of wear, sliding of the mating surfaces takes place by shear in the liquid oil itself. Coefficients of friction under hydrodynamic lubrication are reduced to 0.001 to 0.005.

Hydrodynamic lubrication is all very well when it functions properly. But real bearings contain dirt—hard particles of silica, usually—and new automobile engines are notorious for containing hard cast-iron dust from machining operations on the engine block. Then, if the particles are thicker than the oil film at its thinnest—abrasive wear will take place. There are two ways of solving this problem. One is to make the mating surfaces harder than the dirt particles. Crankshaft journals are "case-hardened" by special chemical and heat treatments to increase the hardness of the *surface* of the journals to the level at which the dirt is abraded by the journal. (It is important not to harden the *whole* shaft because this will make it brittle and it might then break under shock loading.) However, the bearing surfaces are not hardened in this way. As we shall explain in a minute, it is useful to have a *soft* bearing material. If the bearing metal is soft enough, dirt particles will be pushed into the surface of the bearing and will be taken largely out of harm's way. This property of bearing material is called *embeddability*.

Why is it an advantage to have soft bearing materials? Well, a bearing only operates under conditions of hydrodynamic lubrication when the rotational speed of the journal is high enough. When starting an engine up, or running slowly under high load, hydrodynamic lubrication is not present, and we have to fall back on *boundary lubrication* (see the last chapter). Under these conditions some contact, and wear, of the mating surfaces will occur (this is why car engines last less well when used for short runs rather than long ones). Now crankshafts are difficult and expensive to replace when worn, whereas bearings can be designed to be cheap and easy to replace as shown on Fig. 26.2. It is thus good practice to concentrate as much of the wear as possible on the *bearing*—and, as we showed in our section on *adhesive wear* in the last chapter, this is done by having a soft bearing material: lead, tin, zinc or alloys of these metals.

Now for the snag of a soft bearing material—will it not fail to support the normal operating forces imposed on it by the crankshaft? If nothing special were done to the

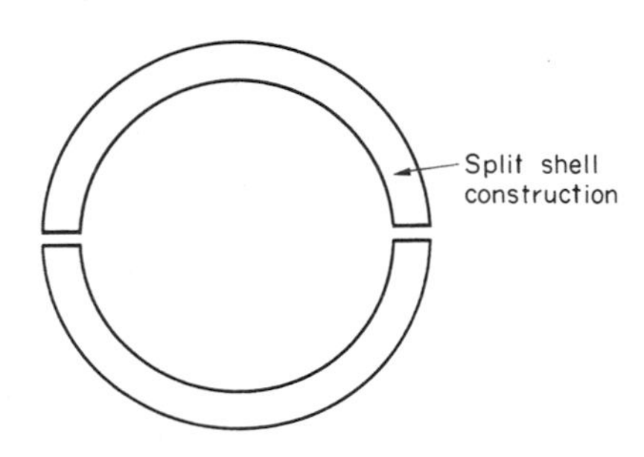

Fig. 26.2. Easily replaceable bearing shells.

material, it would deform under the imposed forces like putty. In practice, by making the layer of soft material *thin*, this difficulty can be avoided.

The way it works is this. If you squeeze a slug of plasticine between two blocks of wood, the slug deforms easily at first, but as the plasticine layer gets thinner and thinner, more and more lateral flow is needed to make it spread, and the pressure required to cause this flow gets bigger and bigger. The plasticine is constrained by the blocks so that it can never be squeezed out altogether—that would take an infinite pressure. This principle of *plastic constraint* is used in bearing design by depositing a very thin layer (about 0.03 mm thick) of soft alloy on to the bearing shell. This is thick enough to embed most dirt particles, but thin enough to support the journal forces.

This soft bearing material also has an important role to play if there is a failure in the oil supply to the bearing. In this case, frictional heating will rapidly increase the bearing temperature, and would normally lead to pronounced metal-to-metal contact, gross atomic bonding between journal and bearing, and seizure. The soft bearing material of low melting point will be able to shear in response to the applied forces, and may also melt locally. This helps protect the journal from severe surface damage, and also helps to avoid component breakages that might result from sudden locking of mating surfaces.

The third advantage of a soft bearing material is *conformability*. Slight misalignments of bearings can be self-correcting if plastic flow occurs easily in the bearing metal (Fig. 26.3). Clearly there is a compromise between load-bearing ability and conformability.

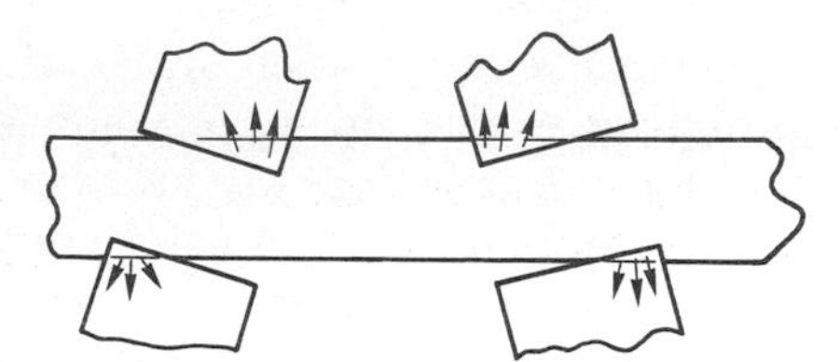

Fig. 26.3. Conformability of bearings; a conformable bearing material will flow to adjust to minor misalignments.

Because our thin overlay of lead–tin can get worn away under severe operating conditions before the end of the normal life of the bearing, it is customary to put a second thicker, and therefore harder, layer between the overlay and the steel backing strip (Fig. 26.4). The alloys normally used are copper–lead, or aluminium–tin. In the event of the wearing through of the overlay they are still soft enough to act as bearing materials without immediate damage to the journal.

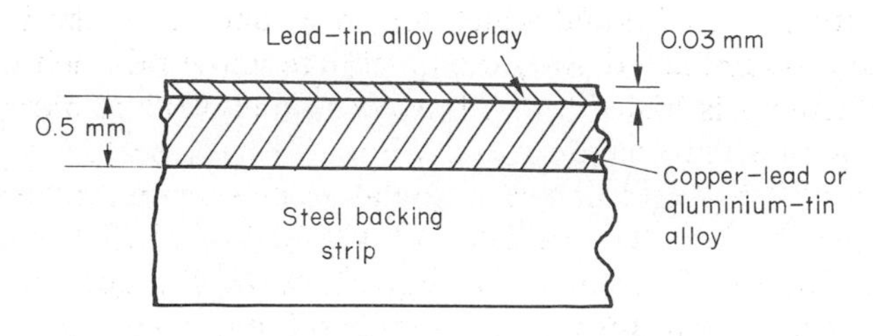

Fig. 26.4. A schematic cross-section through a typical layered bearing shell.

CASE STUDY 2: MATERIALS FOR SKIS AND SLEDGE RUNNERS

Skis, both for people and for aircraft, used to be made of waxed wood. Down to about −10°C, the friction of waxed wood on snow is very low—μ is about 0.02—and if this were not so, planes could not take off from packed-snow runways, and the winter tourist traffic to Switzerland would drop sharply. Below −10°C, bad things start to happen (Fig. 26.5): μ rises sharply to about 0.4. Polar explorers have observed this

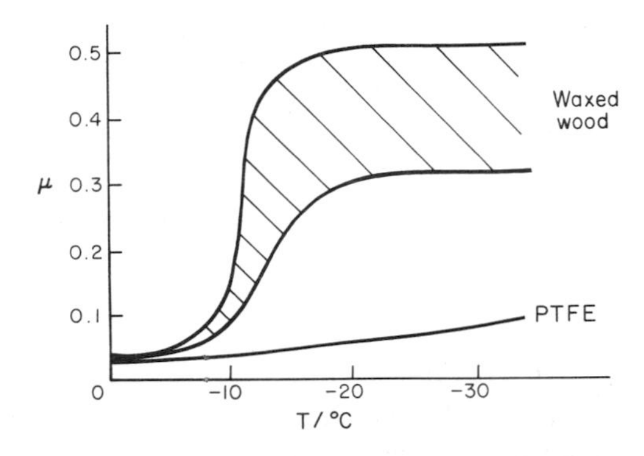

Fig. 26.5. Friction of materials on ice at various temperatures.

repeatedly. Wright, a member of the 1911–13 Scott expedition, writes: "Below 0°F (−18°C) the friction (on the sledge runners) seemed to increase progressively as the temperature fell"; it caused the expedition considerable hardship. What determines the friction of skis on snow?

Ice differs from most materials in that its melting point *drops* if you compress it. It is widely held that pressure from the skis causes the snow beneath to melt, but this is nonsense: the pressure of a large man distributed over a ski lowers the melting point of ice by about 0.0001°C and, even if his weight is carried by asperities which touch the ski over only 10^{-3} of its nominal area, the depression of the melting point is still only 0.1°C. Pressure melting, then, cannot account for the low friction shown in Fig. 26.5. But as the large man starts to descend the ski slope, work is done against the frictional forces, and heat is generated at the sliding surface. This heat is sufficient to melt a layer of ice, producing a thin film of water, at points where asperities touch the ski: the man hydroplanes along on a layer of water generated by his own friction. The principle is exactly like that of the lead–bronze bearing, in which local hot spots melt the lead, producing a lubricating film of liquid which lowers μ and saves the bearing.

Below −10°C, heat is conducted away too quickly to allow this melting—and because their thermal conductivity is high, skis with exposed metal (aluminium or steel edges) are slower at low temperatures than those without. At these low temperatures, the mechanism of friction is the same as that of metals: ice asperities adhere to the ski and must be sheared when it slides. The value of μ (0.4) is close to that calculated from the shearing model in Chapter 25. This is a large value of the coefficient of friction—enough to make it very difficult for a plane to take off, and increasing by a factor of 10 the work required to pull a loaded sledge. What can be done to reduce it?

This is a standard friction problem. A glance at Table 25.1 shows that, when ceramics slide on polymers, μ can be as low as 0.04. Among the polymers with the lowest coefficients are PTFE ("Teflon") and polyethylene. By coating the ski or sledge runners with these materials, the coefficient of friction stays low, *even* when the temperature is so low that frictional heating is unable to produce a boundary layer of water (Fig. 26.5). Aircraft and sports skis now have polyethylene or Teflon undersurfaces; the Olympic Committee has banned their use on bob-sleds, which already, some think, go fast enough.

Case study 3: high-friction rubber

So far we have talked of ways of reducing friction. But for many applications—brake pads, clutch linings, climbing boots, and above all, car tyres—we want as much friction as we can get.

The frictional behaviour of rubber is quite different from that of metals. In Chapter 25 we showed that when metallic surfaces were pressed together, the bulk of the deformation at the points of contact was plastic; and that the friction between the surfaces arose from the forces needed to shear the junctions at the areas of contact.

But rubber deforms *elastically* up to very large strains. When we bring rubber into contact with a surface, therefore, the deformation at the contact points is *elastic*. These elastic forces still squeeze the atoms together at the areas of contact, of course; adhesion will still take place there, and shearing will still be necessary if the surfaces are to slide. This is why car tyres grip well in dry conditions. In *wet* conditions, the situation is different; a thin lubricating film of water and mud forms between rubber and road, and this will shear at a stress a good deal lower than previously with dangerous consequences. Under these circumstances, another mechanism of friction operates to help prevent a skid.

All roads have a fairly rough surface. The high spots push into the tyre, causing a considerable local elastic deformation (Fig. 26.6). As the tyre skids, it slips forward

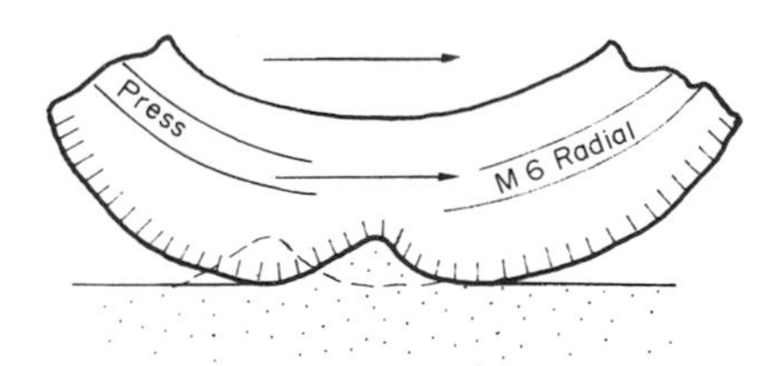

Fig. 26.6. Skidding on a rough road surface deforms the tyre material elastically.

over the rough spots. The region of rubber that was elastically deformed by the rough spot now relaxes, whilst the rubber just behind this region becomes compressed as it reaches the rough spot. Now, all rubbers exhibit some *anelasticity* (Chapter 8); the stress–strain curve looks like Fig. 26.7. As the rubber is compressed, work is done on it equal to the area under the upper curve; but if the stress is removed we do not get all this work back. Part of it is dissipated as heat—the part shown as the shaded area

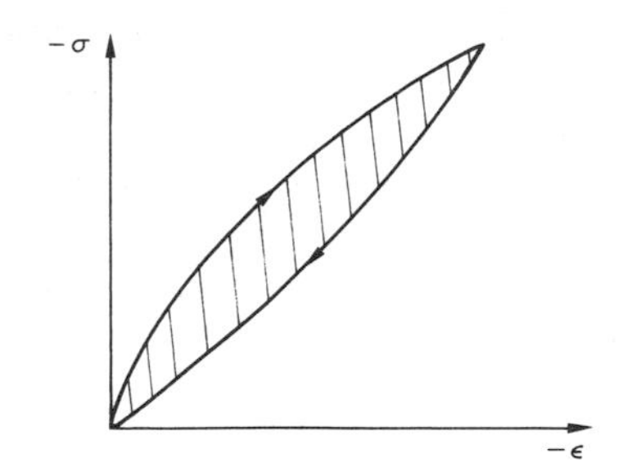

Fig. 26.7. Work is needed to cycle rubber elastically.

between the loading and the unloading curve. So to make the tyre slide on a rough road we have to do work, even when the tyre is well lubricated. Special rubbers have been developed for high-loss characteristics (called "high-loss" or "high-hysteresis" rubbers) and these have excellent skid resistance even in wet conditions.

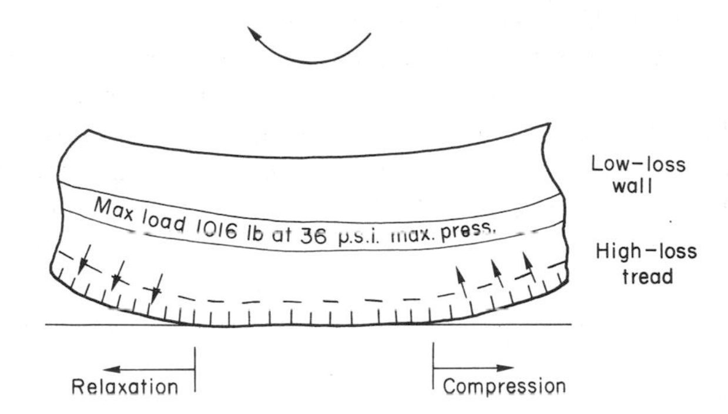

Fig. 26.8. Anti-skid tyres, with a high-loss tread (for maximum grip) and a low-loss wall (for minimum heating up).

There is one obvious drawback of high-hysteresis rubber. In normal rolling operation, considerable elastic deformations still take place in the tyre wall, and high-loss tyres will generate considerable heat. The way out is to use a low-loss tyre covered with a high-loss tread—another example of design using composite materials (Fig. 26.8).

Further reading

F. P. Bowden and D. Tabor, *Friction—An Introduction to Tribology*, Heinemann Science Study Series, No. 41, 1974.

F. P. Bowden and D. Tabor, *The Friction and Lubrication of Solids*, Oxford University Press, Part I, 1950; Part II, 1965.

A. D. Sarkar, *Wear of Metals*, Pergamon Press, 1976.

P. G. Forrester, "Bearing Materials", *Metallurgical Reviews*, **5,** 507 (1960).

E. R. Braithwaite (ed.), *Lubrication and Lubricants*, Elsevier, 1967.

Final case study

CHAPTER 27

MATERIALS AND ENERGY IN CAR DESIGN

Introduction

The rising cost of energy is now making it a primary factor in design. Economy has long been a factor in aircraft design—it has resulted in the development of whole new ranges of materials over the last 30 years. But now the cost of energy has reached the point that even the very conservative car industry is considering drastic changes in design and material selection in order to build cars that consume less energy.

Energy and cars

Energy is used to build a car, and energy is used to run it. Rising oil prices mean that in 1980, for the first time, the cost of the petrol consumed during the average life of a car is comparable to the cost of the car itself. Consumers, therefore, now want more economical cars. But the real reason for wanting such economies is shown in Table 27.1: 15% of all the energy used in a developed country is consumed by private cars.

TABLE 27.1

Energy to produce cars, per year	= 0.8% to 1.5% of total energy consumed by nation
Energy to move cars, per year	= 15% of total energy consumed by nation
(Transportation of people and goods, total)	= 24% of total energy consumed by nation

The dependence of most developed countries on imported oil is such a liability that they are seeking ways of reducing it. Private transport is an attractive target because to limit its energy consumption does not necessarily depress the economy. In the US, for example, legislation has now been passed requiring that the *average* fleet mileage of a manufacturer increase from 22.5 to 34.5 miles per gallon by 1985. How can this (or equivalent) economy be achieved?

Ways of achieving energy economy

It is clear from Table 27.1 that the energy content of the car itself—that is of the steel, rubber, glass and of the manufacturing process itself—is small: one-twentieth to one-tenth of that required to move the car. This means that there is little point in trying to save energy here; indeed (as we shall see) it may pay to use more energy to make the car (using, for instance, aluminium instead of steel) if this reduces the fuel consumption.

We must focus, then, on reducing the energy used to move the car. There are two routes.

(a) *Improve engine efficiency.* Engines are already remarkably efficient; there is a limit to the economy that can be achieved here, though it can help.

(b) *Reduce the weight of the car.* Figure 27.1 shows how the fuel consumption (g.p.m.) and the mileage (m.p.g.) vary with car weight. There is a linear correlation: halving the weight halves the g.p.m. This is why small cars are more economical than big ones: engine size and performance have some influence, but it is mainly the weight that determines the fuel consumption.

We can, then, reduce the size of cars, but the consumer does not like that. Or we can reduce the *weight* of the car by substituting lighter materials for those used now. It is this last route that many manufacturers are now exploring.

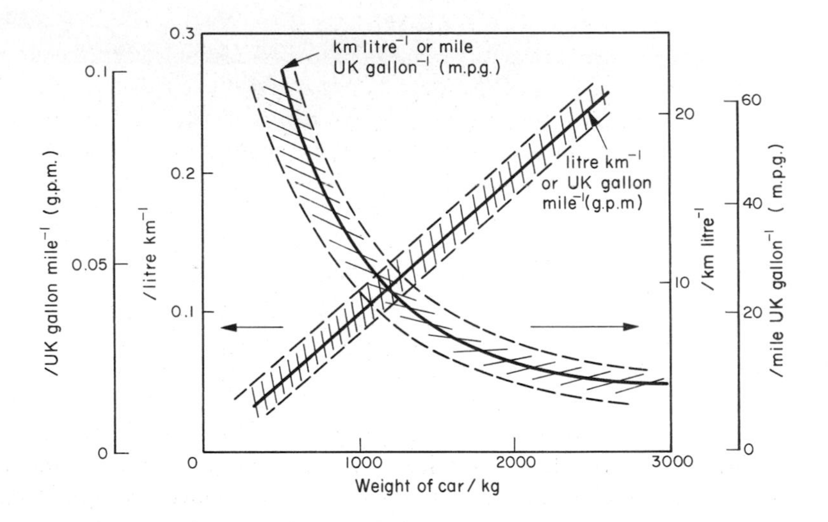

Fig. 27.1. Fuel consumption of production cars.

Material content of a car

As Fig. 27.1 shows, car weight varies from about 400 kg to 2500 kg. In a typical modern production car (Fig. 27.2) this is made up as shown in Table 27.2.

Fig. 27.2. The Volkswagen Passat—a typical 1970s pressed-steel body with no separate chassis. For a *given material* this "monocoque" construction gives a minimum weight-to-strength ratio.

TABLE 27.2

71% Steel:	body shell, panels
15% Cast iron:	engine block; gear box; rear axle
4% Rubber:	tyres; hoses
Balance glass, zinc, copper, aluminium, polymers	

Alternative materials

Primary mechanical properties

Candidate materials for substitutes must be lighter than steel, but structurally equivalent. For the engine block, the choice is an obvious one: aluminium (density $2.7\ \mathrm{Mg\,m^{-3}}$) or possibly magnesium (density $1.8\ \mathrm{Mg\,m^{-3}}$) replace an *equal volume* of cast iron (density $7.7\ \mathrm{Mg\,m^{-3}}$) with an immediate weight reduction on this component of 2.8 to 4.3 times. The production methods remain almost unchanged. Many manufacturers have made this change already.

The biggest potential weight saving, however, is in the body panels, which make up 60% of the weight of the vehicle. Here the choice is more difficult. Candidate materials are given in Table 27.3.

How do we assess the possible weight saving? Just replacing a steel panel with an aluminium-alloy or fibreglass one of equal thickness (giving a weight saving which scales as the density) is unrealistic: both the possible substitutes have much lower

TABLE 27.3

Material*	Density ρ/Mg m^{-3}	Young's modulus, E/GN m^{-2}	Yield strength, σ_y/MN m^{-2}	$(\rho/E^{1/3})$ /N$^{2/3}$ m$^{-10/3}$ s^2	$(\rho/\sigma y^{1/2})$ /N$^{1/2}$ m^{-3} s^2
Mild steel	7.8	207	220	1.32	0.53
High-strength steel	7.8	207	up to 500	1.32	0.35
Aluminium alloy	2.7	69	193	0.66	0.19
GFRP (chopped fibre, moulding grade)	1.8	15	75	0.73	0.21

* In the past *wood* was extensively used in car bodies (Fig. 27.3); we will not consider it here because of the difficulty of selecting high-quality timber for mass-production methods.

moduli, and thus will deflect (under given loads) far more; and one of them has a much lower yield strength, and might undergo plastic flow. We need an analysis like that of Chapters 7 and 12.

If (as with body panels) *elastic* deflection is what counts, the logical comparison is for a panel of equal *stiffness*. And if, instead, it is resistance to *plastic* flow which counts (as with bumpers) then the proper thing to do is to compare sections with equal resistance to plastic flow.

Fig. 27.3. The Morris Traveller—a classic of the 1950s with wooden members used as an integral part of the monocoque body shell.

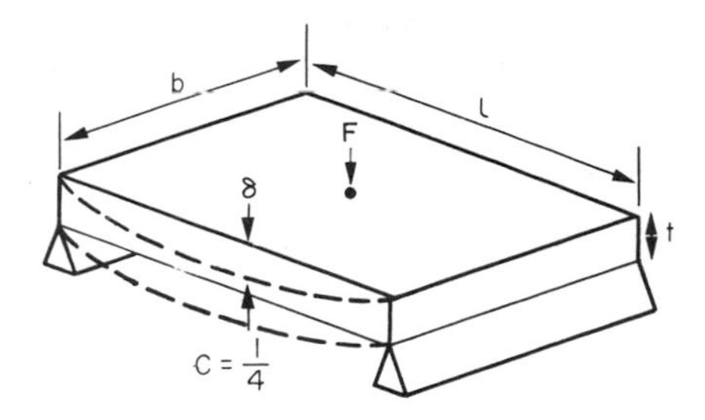

Fig. 27.4. Elastic deflection of a car-body panel.

The *elastic* deflection δ of a panel under a force F (Fig. 27.4) is given by

$$\delta = \frac{Cl^3F}{Ebt^3} \tag{27.1}$$

where t is the sheet thickness, l the length and b the width of the panel. The constant C depends on how the panel is held: at two edges, at four edges, and so on—it does not affect our choice. The mass of the panel is

$$M = \rho btl. \tag{27.2}$$

The quantities b and l are determined by the design (they are the dimensions of the door, trunk lid, etc.). The only variable in controlling the stiffness is t. Now from eqn. (27.1)

$$t = \left(\frac{Cl^3F}{\delta Eb}\right)^{1/3}. \tag{27.3}$$

Substituting this expression for t in eqn. (27.2) gives, for the mass of the panel,

$$M = \left\{C\left(\frac{F}{\delta}\right)l^6b^2\right\}^{1/3}\left(\frac{\rho}{E^{1/3}}\right). \tag{27.4}$$

Thus, for a given *stiffness* (F/δ) and *panel* (l, b, C) the lightest panel is the one with the smallest value of $\rho/E^{1/3}$.

We can perform a similar analysis for *plastic* yielding. A panel with the section shown in Fig. 27.5 yields at a load

$$F = \left(\frac{Cbt^2}{l}\right)\sigma_y. \tag{27.5}$$

As before, the panel mass is

$$M = \rho btl \tag{27.6}$$

with the only variable t given by

$$t = \left(\frac{Fl}{Cb\sigma_y}\right)^{1/2} \tag{27.7}$$

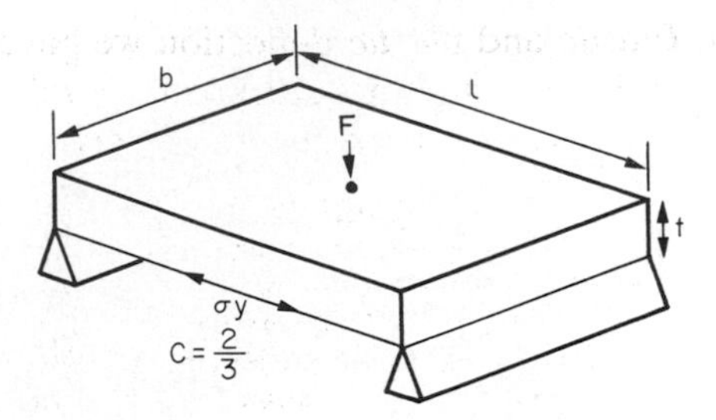

Fig. 27.5. Yielding of a car-body panel.

from eqn. (27.5). Substituting for t in eqn. (27.6) gives

$$M = \left(\frac{Fbl^3}{C}\right)^{1/2}\left(\frac{\rho}{\sigma_y^{1/2}}\right). \tag{27.8}$$

The panel with the smallest value of $\rho/\sigma_y^{1/2}$ is the one we want.

We can now assess candidate materials more sensibly on the basis of the data given in Table 27.3. For the majority of body panels (for which *elastic* deflection determines design) high-strength steel offers no advantage—though its use in bumpers, front and rear header panels, engine mounts, bulkheads, and so forth, makes sense, with a weight saving ($\rho/\sigma_y^{1/2}$) of up to 1.8 times. Both aluminium alloy and fibreglass offer real weight savings of up to 2 times on body panels ($\rho/E^{1/3}$), and up to 3 times ($\rho/\sigma_y^{1/2}$) on bumpers, headers, bulkheads, etc. This makes possible a saving of at least 30% on the weight of the vehicle; if, in addition, an aluminium engine block is used, the overall weight saving is larger still. These are very substantial savings—sufficient to achieve the desired increase in mileage per gallon from 22.5 to 34.5 without any decrease in the size of the car, or increase in engine efficiency. So they are obviously worth examining more closely. What, then, of the *other* properties required of the substitute materials?

Secondary properties

Although resistance to deflection and plastic yielding are obviously of first importance in choosing alternative materials, other properties enter into the selection. Let us look at these briefly. Table 27.4 lists the conditions imposed by the service environment.

TABLE 27.4
SERVICE ENVIRONMENT

Loading	Static →	Elastic or plastic deflection	
	Impact →	Elastic or plastic deflection	
	Impact →	Fracture	
	Fatigue →	Fatigue fracture	
	Long-term static →	Creep	
Physical environment	$-40°C < T < 120°C$		
	55% < relative humidity < 100%		
Chemical environment		Water	Petrol
		Oil	Antifreeze
		Brake fluid	Salt
		Transmission fluid	

Consider these in turn. *Elastic* and *plastic* deflection we have dealt with already. The toughness of steel is so high that *fracture* of a steel panel is seldom a problem. But what about the other materials? The data for toughness are given in Table 27.5.

TABLE 27.5
TOUGHNESS, FATIGUE, CREEP

Material	Toughness, G_c/kJ m^{-2}	Tolerable crack length /mm	Fatigue	Creep
Mild steel	≈100	≈140	O.K.	O.K.
High-strength steel	≈100	≈26	O.K.	O.K.
Aluminium alloy	≈20	≈12	O.K.	O.K.
GFRP (chopped fibre, moulding grade)	≈37	≈30	O.K.	Creep above 60°C

But what is the proper way to use toughness values? The most sensible thing to do is ask: suppose the panel is loaded up to its yield load (above this load we *know* it will begin to fail—by plastic flow—so it does not matter whether other failure mechanisms also appear); what is the maximum crack size that is still stable? If this is large enough that it should not appear in service, we are satisfied; if not, we must increase the section. This crack size is given (Chapter 13) by

$$\sigma_y\sqrt{\pi a} = K_c = \sqrt{EG_c}$$

from which

$$a_{max} = \frac{EG_c}{\pi\sigma_y^2}.$$

The resulting crack lengths are given in Table 27.5. A panel with a crack longer than this will fail by "tearing"; with one less than this will simply fail by general yield, i.e. it will bend permanently. Though the tolerable crack lengths are much shorter in replacement materials than in steel, they are still large enough to permit the replacement materials to be used.

Fatigue (Chapter 15) is always a potential problem with any structure subject to varying loads: anything from the loading due to closing the door to that caused by engine vibration can, potentially, lead to failure. Nevertheless, the fatigue strength of all these materials is adequate.

Creep (Chapter 17) is not normally a problem a designer considers when designing a car with metals: the maximum service temperature reached is 120°C (panels near the engine, under extreme conditions), and neither steel nor aluminium alloys creep significantly at these temperatures. But GFRP does. Above 60°C creep-rates are significant. GFRP shows a classic three-stage creep curve, ending in failure; so that extra reinforcement or heavier sections will be necessary where temperatures exceed this value.

More important than either creep or fatigue in current car design is the *effect of environment* (Chapter 23). An appreciable part of the cost of a new car is contributed by the manufacturing processes designed to prevent rusting; and these processes only partly work—it is body-rust that ultimately kills a car, since the mechanical parts (engine, etc.) can be replaced quite easily, as often as you like.

Steel is particularly bad in this regard. In ordinary circumstances, aluminium is much better as we showed in the chapters on corrosion. Although the effect of salt on aluminium is bad, heavy anodising will slow down even that form of attack to tolerable levels (the masts of most modern yachts are made of anodised aluminium alloy, for example).

So aluminium alloy is good: it resists all the fluids likely to come in contact with it. What about GFRP? The strength of GFRP is reduced by up to 20% by continuous immersion in most of the fluids—even salt water—with which it is likely to come into contact; but (as we know from fibreglass boats) this drop in strength is not critical, and it occurs without visible corrosion, or loss of section. In fact, GFRP is much more corrosion-resistant, in the normal sense of loss-of-section, than steel.

Production methods

The biggest penalty one has to pay in switching materials is likely to be the higher production costs. *High-strength steel,* of course, presents almost no problem. The yield strength is higher, but the section is thinner, so that only slight changes in punches, dies and presses are necessary, and once these are paid for, the extra cost is merely that of the material.

At first sight, the same is true of *aluminium alloys.* But because they are heavily alloyed (to give a high yield strength) their *ductility* is low. If expense is unimportant, this does not matter; some early Rolls-Royce cars (Fig. 27.6) had aluminium bodies which were formed into intricate shapes by laborious hand-beating methods, with

Fig. 27.6. A 1932 Rolls-Royce. Mounted on a separate steel chassis is an all-aluminium hand-beaten body by the famous coachbuilding firm of James Mulliner. Any weight advantage due to the use of aluminium is totally outweighed by the poor weight-to-strength ratio of separate-chassis construction; but the bodywork remains immaculate after 48 years of continuous use!

frequent annealing of the aluminium to restore its ductility. But in mass production we should like to deep draw body panels in one operation—and then low ductility is much more serious. The result is a loss of design flexibility: there are more constraints on the use of aluminium alloys than on steel; and it is this, rather than the cost, which is the greatest obstacle to the wholesale use of aluminium in cars.

GFRP looks as if it would present production problems: you may be familiar with the tedious hand lay-up process required to make a fibreglass boat or canoe. But mass-production methods have now been developed to handle GFRP, and quite a few modern cars now have GFRP bodies (Fig. 27.7; although, in most, there is a separate steel chassis to carry much of the load, and realistic weight savings will only be made if the *whole* load-bearing structure is made from GFRP). In producing GFRP car panels a slug of polyester resin, with *chopped* glass fibres mixed in with it, is dropped into a heated split mould (Fig. 27.8). As the polyester used is a *thermoset* it will "go off" in the hot mould, after which the solid moulding can be ejected. Modern methods allow a press like this to produce one moulding per minute—still slower than steel pressing, but practical. Moulding (as this is called) brings certain advantages. It offers great design flexibility—particularly in change of section, and sharp detail—which cannot be achieved with steel. And GFRP mouldings often result in consolidation of components, reducing assembly costs.

Fig. 27.7. A 1979 Lotus Elite, with a GFRP body (but still mounted on a steel chassis—which does not give anything like the weight saving expected with an all-GFRP monocoque structure).

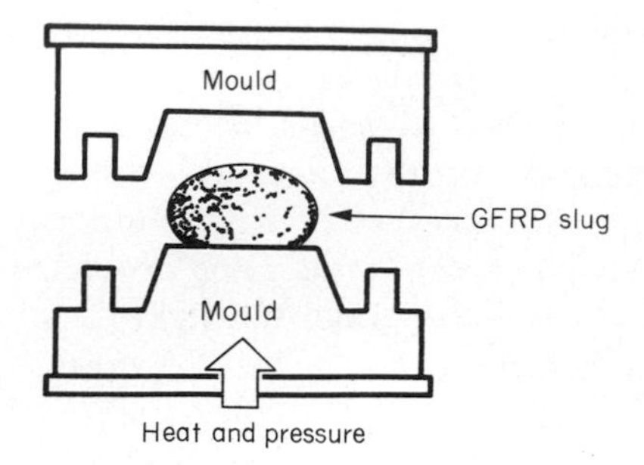

Fig. 27.8. Compression moulding of car-body components.

Conclusions

The conclusions are set out in the table below.

A. HIGH-STRENGTH STEEL

For	Against
Retains all existing technology	Weight saving only appreciable in designing against *plastic* flow

Use in selected applications, e.g. bumpers.

B. ALUMINIUM ALLOY

For	Against
Large weight saving in both body shell and engine block	Unit cost higher
Retains much existing technology	Deep drawing properties poor—loss in design flexibility
Corrosion resistance excellent	

Aluminium alloy offers saving of up to 40% in total car weight, but at increased unit cost. Good short-term solution.

C. GFRP

For	Against
Large weight saving in body shell	Unit cost higher
Corrosion resistance excellent	Massive changes in manufacturing technology
Great gain in design flexibility and some parts consolidation	Designer must cope with some creep

GFRP offers savings of up to 30% in total car weight, at some increase in unit cost and considerable capital investment in new equipment. Best long-term solution.

APPENDIX 1

EXAMPLES

1. (a) Commodity A is currently consumed at the rate C_A tonnes per year, and commodity B at the rate C_B tonnes per year ($C_A > C_B$). If the two consumption rates are increasing exponentially to give growths in consumption after each year of r_A% and r_B%, respectively ($r_A < r_B$), derive an equation for the time, measured from the present day, before the annual consumption of B exceeds that of A.

(b) The table shows 1980 figures for consumptions and growth rates of steel, aluminium and plastics. What are the doubling times (in years) for consumption of these commodities?

(c) Calculate the number of years, measured from 1980, before the consumption of (a) aluminium and (b) polymers would exceed that of steel, *if exponential growth continued.* Is this continued growth probable?

Material	Current (1980) world consumption /tonnes year^{-1}	Projected growth rate in consumption /% year^{-1} (1980)
Iron and Steel	5.5×10^8	1.8
Aluminium	1.9×10^7	6.4
Polymers	1.1×10^8	12

Answers:

(a) $t = \dfrac{100}{r_B - r_A} \ln \left(\dfrac{C_A}{C_B}\right)$.

(b) Doubling times: steel, 38.5 years; aluminium, 10.8 years; plastics, 5.8 years.

(c) If exponential growth continued, aluminium would overtake steel in 73 years (A.D. 2053); polymers would overtake steel in 16 years (A.D. 1996).

2. (a) Discuss ways of conserving engineering materials, and the technical and social problems involved in implementing them.

(b) 12% of the world production of lead is used dissipatively as an antiknock compound in petrol. If laws were passed to prevent this use, how many years would it require before the consumption of lead returned to the level obtaining *just* before the new laws took effect? Assume that the other uses of lead continue to grow at an average rate of 2% per year.

Answer: (b) 6.4 years.

3. (a) Explain what is meant by *exponential growth* in the consumption of a material.

(b) A material is consumed at C_0 tonne year^{-1} in 1980. Consumption in 1980 is increasing at r% year^{-1}. If the resource base of the material is Q tonnes, and consumption *continues* to increase at r% year^{-1}, show that the resource will be half exhausted after a time, $t_{1/2}$, given by

$$t_{1/2} = \frac{100}{r} \ln \left\{ \frac{rQ}{200C_0} + 1 \right\}.$$

(c) Discuss, giving specific examples, the factors that might cause a decrease in the rate of consumption of a potentially scarce material.

4. Use the information given in Table 2.1 (Prices of Materials) and in Table 2.5 (Energy Content of Materials) to calculate the approximate cost of (a) aluminium, (b) low-density polyethylene, (c) mild steel and (d) cement in 1990, assuming that oil increases in price by a factor of 4 and that labour and other manufacturing costs increase by a factor of 2 between 1980 and 1990.

Comment on the implications of your results (e.g. Which commodities have increased by the largest factor? How have the relative costs of materials changed? What are the implications for the use of polymers?).

Answers: (a) aluminium, UK£3000 (US\$6600) tonne^{-1}; (b) polyethylene, UK£1500 (US\$3300) tonne^{-1}; (c) mild steel, UK£600 (US\$1320) tonne^{-1}; (d) cement, UK£80 (US\$176) tonne^{-1}.

5. (a) Define *Poisson's ratio*, ν, and the *dilatation*, Δ, in the straining of an elastic solid.

(b) Calculate the dilatation Δ in the uniaxial elastic extension of a bar of material, assuming strains are small, in terms of ν and the tensile strain, ε. Hence find the value of ν for which the volume change during elastic deformation is zero.

(c) Poisson's ratio for most metals is about 0.3. For cork it is close to zero; for rubber it is close to 0.5. What are the approximate volume changes in each of these materials during an elastic tensile strain of ε?

Answers: (b) 0.5, (c) "most metals": 0.4ε; cork: ε; rubber: 0.

6. The potential energy U of two atoms, a distance r apart, is

$$U = -\frac{A}{r^m} + \frac{B}{r^n}, \quad m = 2, \quad n = 10.$$

Given that the atoms form a stable molecule at a separation of 0.3 nm with an energy of -4 eV, calculate A and B. Also find the force required to break the molecule, and the critical separation at which the molecule breaks. You should sketch an energy/distance curve for the atom, and sketch beneath this curve the appropriate *force*/distance curve.

Answers: A: 7.2×10^{-20} J nm^2; B: 9.4×10^{-25} J nm^{10}; Force: 2.39×10^{-9} N at 0.352 nm.

7. The potential energy U of a pair of atoms in a solid can be written as

$$U = \frac{-A}{r^m} + \frac{B}{r^n}$$

where r is the separation of the atoms, and A, B, m and n are positive constants. Indicate the physical significance of the two terms in this equation.

A material has a cubic unit cell with atoms placed at the corners of the cubes. Show that, when the material is stretched in a direction parallel to one of the cube edges, Young's modulus E is given by

$$E = \frac{mnkT_M}{\Omega}$$

where Ω is the mean atomic volume, k is Boltzmann's constant and T_M is the absolute melting temperature of the solid. You may assume that $U(r_0) = -kT_M$, where r_0 is the equilibrium separation of a pair of atoms.

8. The table below gives the Young's modulus, E, the atomic volume, Ω, and the melting temperature, T_M, for a number of metals. If

$$E \simeq \frac{\tilde{A}kT_M}{\Omega}$$

(where k is Boltzmann's constant and $\tilde{A}$ is a constant), calculate and tabulate the value of the constant $\tilde{A}$ for each metal. Hence find an arithmetic mean of $\tilde{A}$ for these metals.

Use the equation, with the average $\tilde{A}$, to calculate the approximate Young's modulus of (a) diamond and (b) ice. Compare these with the experimental values of 1.0×10^{12} N m^{-2} and 7.7×10^{9} N m^{-2}, respectively. Watch the units!

Material	$\Omega \times 10^{29}$/m^3	T_M/K	E/GN m^{-2}
Nickel	1.09	1726	214
Copper	1.18	1356	124
Silver	1.71	1234	76
Aluminium	1.66	933	69
Lead	3.03	600	14
Iron	1.18	1753	196
Vanadium	1.40	2173	130
Chromium	1.20	2163	289
Niobium	1.80	2741	100
Molybdenum	1.53	2883	360
Tantalum	1.80	3271	180
Tungsten	1.59	3683	406

Data for ice and for diamond.

Ice	Diamond
$\Omega = 3.27 \times 10^{-29}$ m^3	$\Omega = 5.68 \times 10^{-30}$ m^3
$T_M = 273$ K	$T_M = 4200$ K
$E = 7.7 \times 10^{9}$ N m^{-2}	$E = 1.0 \times 10^{12}$ N m^{-2}

Answers: Mean $\tilde{A} = 88$. Calculated moduli: diamond, 9.0×10^{11} N m^{-2}; ice, 1.0×10^{10} N m^{-2}.

9. (a) Calculate the density of an f.c.c. packing of spheres of unit density.

(b) If these same spheres are packed to form a *glassy* structure, the arrangement is called "dense random packing" and has a density of 0.636. If crystalline f.c.c. nickel has a density of 8.90 Mg m^{-3}, calculate the density of glassy nickel.

Answers: (a) 0.740, (b) 7.65 Mg m^{-3}.

10. (a) Sketch three-dimensional views of the unit cell of a b.c.c. crystal, showing a (100) plane, a (110) plane, a (111) plane and a (210) plane.

(b) The slip planes of b.c.c. iron are the {110} planes: sketch the atom arrangement in these planes, and mark the ⟨111⟩ slip directions.

(c) Sketch three-dimensional views of the unit cell of an f.c.c. crystal, showing a [100], a [110], a [111] and a [211] direction.

(d) The slip planes of f.c.c. copper are the {111} planes: sketch the atom arrangement in these planes and mark the ⟨110⟩ slip directions.

11. (a) The atomic diameter of an atom of nickel is 0.2492 nm. Calculate the lattice constant a of f.c.c. nickel.

(b) The atomic weight of nickel is 58.71 kg kmol^{-1}. Calculate the density of nickel. (Calculate first the mass per atom, and the number of atoms in a unit cell.)

(c) The atomic diameter of an atom of iron is 0.2482 nm. Calculate the lattice constant a of b.c.c. iron.

(d) The atomic weight of iron is 55.85 kg kmol^{-1}. Calculate the density of iron.

Answers: (a) 0.352 nm, (b) 8.91 Mg m^{-3}, (c) 0.287 nm, (d) 7.88 Mg m^{-3}.

12. Crystalline copper and magnesium have face-centred-cubic and close-packed-hexagonal structures respectively.

(a) Assuming that the atoms can be represented as hard spheres, calculate the percentage of the volume occupied by atoms in each material.

(b) Calculate, from first principles, the dimensions of the unit cell in copper and in magnesium.

(The densities of copper and magnesium are 8.96 Mg m^{-3} and 1.74 Mg m^{-3}, respectively.)

Answers: (a) 74% for both; (b) copper: $a = 0.361$ nm; magnesium: $a = 0.320$ nm; $c = 0.523$ nm.

13. The table lists Young's modulus, $E_{composite}$, for a glass-filled epoxy composite. The

Volume fraction of glass, V_f	$E_{composite}$ /GN m^{-2}
0	5.0
0.05	5.5
0.10	6.4
0.15	7.8
0.20	9.5
0.25	11.5
0.30	14.0

material consists of a volume fraction V_f of glass particles (Young's modulus, E_f, 80 GN m^{-2}) dispersed in a matrix of epoxy (Young's modulus, E_m, 5 GN m^{-2}).

Calculate the upper and lower values for the modulus of the composite material, and plot them, together with the data, as a function of V_f. Which set of values most nearly describes the results? Why? How does the modulus of a random chopped-fibre composite differ from those of an aligned continuous-fibre composite?

14. A composite material consists of parallel fibres of Young's modulus E_F in a matrix of Young's modulus E_M. The volume fraction of fibres is V_F. Derive an expression for E_C, Young's modulus of the composite along the direction of the fibres, in terms of E_F, E_M and V_F. Obtain an analogous expression for the density of the composite, ρ_C. Using material parameters given below, find ρ_C and E_C for the following composites: (a) carbon fibre-epoxy resin (V_F=0.5), (b) glass fibre-polyester resin (V_F=0.5), (c) steel-concrete (V_F=0.02).

A uniform, rectangular-section beam of fixed width w, unspecified depth d, and fixed length L rests horizontally on two simple supports at either end of the beam. A concentrated force F acts vertically downwards through the centre of the beam. The deflection, δ, of the loaded point is

$$\delta = \frac{FL^3}{4E_C wd^3}$$

ignoring the deflection due to self weight. Which of the three composites will give the lightest beam for a given force and deflection?

Material	Density /Mg m^{-3}	Young's modulus /GN m^{-2}
Carbon fibre	1.90	390
Glass fibre	2.55	72
Epoxy resin } Polyester resin }	1.15	3
Steel	7.90	200
Concrete	2.40	45

Answers: $E_C = E_F V_F + (1 - V_F) E_M$; $\rho_C = \rho_F V_F + (1 - V_F)\rho_M$.
(a) ρ_C = 1.53 Mg m^{-3}, E_C = 197 GN m^{-2}; (b) ρ_C = 1.85 Mg m^{-3}, E_C = 37.5 GN m^{-2};
(c) ρ_C = 2.51 Mg m^{-3}, E_C = 48.1 GN m^{-2}.
Carbon fibre/Epoxy resin.

15. Indicate, giving specific examples, why some composite materials are particularly attractive in materials applications.

A composite material consists of flat, thin metal plates of uniform thickness glued one to another with a thin, epoxy-resin layer (also of uniform thickness) to form a "multi-decker-sandwich" structure. Young's modulus of the metal is E_1, that of the epoxy resin is E_2 (where $E_2 < E_1$) and the volume fraction of metal is V_1. Find the ratio of the maximum composite modulus to the minimum composite modulus in terms of E_1, E_2 and V_1. Which value of V_1 gives the largest ratio?

Answer: Largest ratio when V_1 = 0.5.

16. (a) Define a high polymer; list three engineering polymers.

(b) Define a thermoplastic and a thermoset.

(c) Distinguish between a glassy polymer, a crystalline polymer and a rubber.

(d) Distinguish between a cross-linked and a non-cross-linked polymer.

(e) What is a co-polymer?

(f) List the monomers of polyethylene (PE), polyvinyl chloride (PVC), and polystyrene (PS).

(g) What is the glass transition temperature, T_G?

(h) Explain the change of moduli of polymers at the glass transition temperature.

(i) What is the order of magnitude of the number of carbon atoms in a single molecule of a high polymer?

(j) What is the range of temperature in which T_G lies for most engineering polymers?

(k) How would you increase the modulus of a polymer?

17. (a) Select the material for the frame of the ultimate bicycle—meaning the lightest possible for a given stiffness. You may assume that the tubes of which the frame is made are cantilever beams (of length l) and that the elastic bending displacement δ of one end of a tubular beam under a force F (the other end being rigidly clamped) is

$$\delta = \frac{Fl^3}{3E\pi r^3 t}.$$

$2r$ is the diameter of the tube (fixed by the designer) and t is the wall thickness of the tube, which you may vary. t is much less than r. Find the combination of material properties which determine the mass of the tube for a given stiffness, and hence make your material selection using data given in Chapters 3 and 5. Try steel, aluminium alloy, wood, GFRP and CFRP.

(b) Which of the following materials leads to the cheapest bicycle frame, for a given stiffness: mild steel, aluminium alloy, titanium alloy, GFRP, CFRP and hard wood?

Answers: (a) CFRP, (b) steel.

18. Explain what is meant by the *ideal strength* of a material. Show how dislocations can allow metals and alloys to deform plastically at stresses that are much less than the ideal strength. Indicate, giving specific examples, the ways in which metals and alloys may be made harder.

19. The energy per unit length of a dislocation is $\frac{1}{2}Gb^2$. Dislocations dilate a close-packed crystal structure very slightly, because at their cores, the atoms are not close-packed: the dilatation is $\frac{1}{4}b^2$ per unit length of dislocation. It is found that the density of a bar of copper changes from 8.9323 Mg m^{-3} to 8.9321 Mg m^{-3} when it is very heavily deformed. Calculate (a) the dislocation density introduced into the copper by the deformation and (b) the energy associated with that density. Compare this result with the latent heat of melting of copper (1833 MJ m^{-3}) (b for copper is 0.256 nm; G is $\frac{3}{8}E$).

Answers: (a) 1.4×10^{15} m^{-2}, (b) 2.1 MJ m^{-3}.

20. Explain briefly what is meant by a *dislocation.* Show with diagrams how the motion of (a) an edge dislocation and (b) a screw dislocation can lead to the plastic deformation of a crystal under an applied shear stress. Show how dislocations can account for the following observations:

(a) cold working makes aluminium harder;
(b) an alloy of 20% Zn, 80% Cu is harder than pure copper;
(c) the hardness of nickel is increased by adding particles of thorium oxide.

21. (a) Derive an expression for the shear stress τ needed to bow a dislocation line into a semicircle between small hard particles a distance L apart.

(b) A polycrystalline aluminium alloy contains a dispersion of hard particles of diameter 10^{-8} m and average centre-to-centre spacing of 6×10^{-8} m measured in the slip planes. Estimate their contribution to the tensile yield strength, σ_y, of the alloy.

(c) The alloy is used for the compressor blades of a small turbine. Adiabatic heating raises the blade temperature to 150°C, and causes the particles to coarsen slowly. After 1000 hours they have grown to a diameter of 3×10^{-8} m and are spaced 18×10^{-8} m apart. Estimate the drop in yield strength. (The shear modulus of aluminium is 26 GN m^{-2}, and $b = 0.286$ nm.)

Answers: (b) 450 MN m^{-2}, (c) 300 MN m^{-2}.

22. Nine strips of pure, fully annealed copper were deformed plastically by being passed between a pair of rotating rollers so that the strips were made thinner and longer. The increases in length produced were 1, 10, 20, 30, 40, 50, 60, 70 and 100% respectively. The diamond-pyramid hardness of each piece was measured after rolling. The results were

Nominal strain	0.01	0.1	0.2	0.3	0.4	0.5	0.6	0.7	1.0
Hardness/MN m^{-2}	423	606	756	870	957	1029	1080	1116	1170

Assuming that a diamond-pyramid hardness test creates a further nominal strain, on average, of 0.08, and that the hardness value is 3.0 times the true stress with this extra strain, construct the curve of *nominal* stress against nominal strain, and find:

(a) the tensile strength of copper;
(b) the strain at which tensile failure commences;
(c) the percentage reduction in cross-sectional area at this strain;
(d) the work required to initiate tensile failure in a cubic metre of annealed copper.

Why can copper survive a much higher extension during rolling than during a tensile test?

Answers: (a) 217 MN m^{-2}, (b) 0.6 approximately, (c) 38%, (d) 109 MJ.

23. (a) If the *true* stress–*true* strain curve for a material is defined by

$$\sigma = A\varepsilon^n, \text{ where } A \text{ and } n \text{ are constants,}$$

find the tensile strength σ_{TS}. (Method: first find the equation of the *nominal* stress–*nominal* strain curve, assuming flow does not localise. Differentiate to find the maximum of this curve. Hence find the strain corresponding to the tensile strength. Use this to find the tensile strength itself.)

(b) For a nickel alloy, $n = 0.2$, and $A = 800$ MN m^{-2}. Evaluate the tensile strength of the alloy. Evaluate the true stress in an alloy specimen loaded to σ_{TS}.

Answers: (a) $\sigma_{TS} = \dfrac{An^n}{e^n}$, (b) $\sigma_{TS} = 475$ MN m^{-2}, $\sigma = 580$ MN m^{-2}.

24. (a) Discuss the assumption that, when a piece of metal is deformed at constant temperature, its volume is unchanged.

(b) A ductile metal wire of uniform cross-section is loaded in tension until it just begins to neck. Assuming that volume is conserved, derive a differential expression relating the *true* stress to the *true* strain at the onset of necking.

(c) The curve of true stress against true strain for the metal wire approximates to

$$\sigma = 350\varepsilon^{0.4} \text{ MN m}^{-2}.$$

Estimate the tensile strength of the wire and the work required to take 1 m^3 of the wire to the point of necking.

Answers: (c) 163 MN m^{-2}, 69.3 MJ.

25. (a) One type of hardness test involves pressing a hard sphere (radius r) into the test material under a fixed load F, and measuring the *depth*, h, to which the sphere sinks into the material, plastically deforming it. Derive an expression for the indentation hardness, H, of the material in terms of h, F and r. Assume $h \ll r$.

(b) The indentation hardness, H, is found to be given by $H \approx 3\sigma_y$ where σ_y is the true yield stress at a nominal plastic strain of 8%. If (as in question 23) the true stress–strain curve of a material is given by

$$\sigma = A\varepsilon^n$$

and $n = 0.2$, calculate the tensile strength of a material for which the indentation hardness is 600 MN m^{-2}. You may assume that $\sigma_{TS} = An^n/e^n$ (see question 23).

Answers: (a) $H = \dfrac{F}{2\pi rh}$, (b) $\sigma_{TS} = 198$ MN m^{-2}.

26.

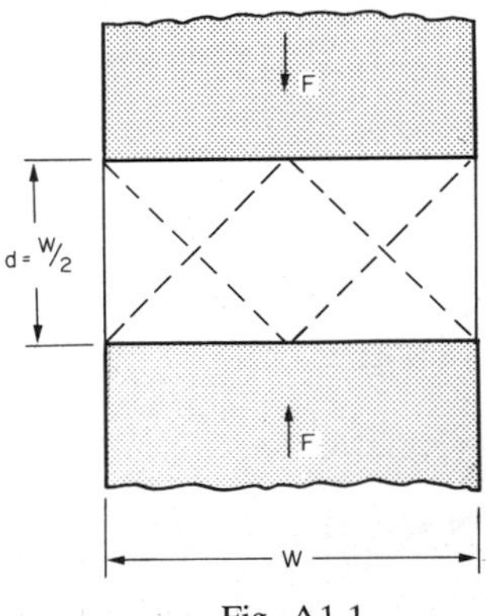

Fig. A1.1.

A metal bar of width w is compressed between two hard anvils as shown in Fig. A1.1. The third dimension of the bar, L, is much greater than w. Plastic deformation takes place as a result of shearing along planes, defined by the dashed lines in the figure, at a shear stress k. Find an upper bound for the load F when (a) there is no friction between anvils and bar, and (b) there is sufficient friction to effectively weld the anvils to the bar. Show that the solution to case (b) satisfies the general formula

$$F \leqslant 2wLk\left(1+\frac{w}{4d}\right)$$

which defines upper bounds for all integral values of $w/2d$.

A composite material used for rock-drilling bits consists of an assemblage of tungsten carbide cubes (each $2\ \mu m$ in size) stuck together with a thin layer of cobalt. The material is required to withstand compressive stresses of $4000\ MN\ m^{-2}$ in service. Use the above equation to estimate an upper limit for the thickness of the cobalt layer. You may assume that the compressive yield stress of tungsten carbide is well above $4000\ MN\ m^{-2}$, and that the cobalt yields in shear at $k = 175\ MN\ m^{-2}$. What assumptions made in the analysis are likely to make your estimate inaccurate?

Answers: (a) $2wLk$, (b) $3wLk$; $0.048\ \mu m$.

27. By calculating the plastic work done in each process, determine whether the bolt

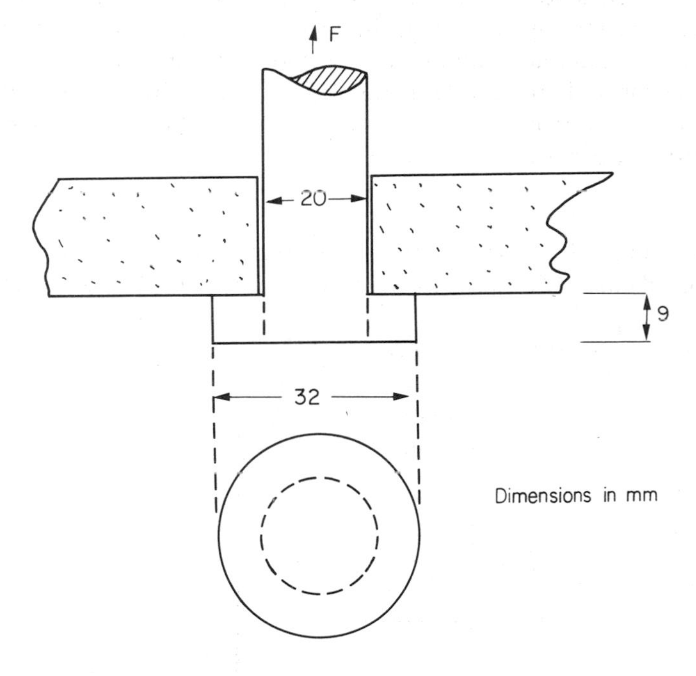

Fig. A1.2.

passing through the plate in Fig. A1.2 will fail, when loaded in tension, by yielding of the shaft or shearing-off of the head. (Assume no work-hardening.)

Answer: The bolt will fail by shearing-off of the head.

28. Sketch curves of the nominal stress against nominal strain obtained from tensile tests on (a) a typical ductile material, (b) a typical non-ductile material. The following data were obtained in a tensile test on a specimen with 50 mm gauge length and a cross-sectional area of 160 mm^2.

Extension/mm	0.050	0.100	0.150	0.200	0.250	0.300	1.25	2.50	3.75	5.00	6.25	7.50
Load/kN	12	25	32	36	40	42	63	80	93	100	101	90

The total elongation of the specimen just before final fracture was 16%, and the reduction in area at the fracture was 64%.

Find the maximum allowable working stress if this is to equal

(a) $0.25 \times$ Tensile strength,

(b) $0.6 \times 0.1\%$ proof stress.

(c) What would have been the elongation and maximum reduction in area if a 150 mm gauge length had been used?

Answers: (a) 160 MN m^{-2}; (b) 131 MN m^{-2}; (c) 12.85%, 64%.

29. A large thick plate of steel is examined by X-ray methods, and found to contain no detectable cracks. The equipment can detect a single edge-crack of depth $a = 1$ mm or greater. The steel has a fracture toughness K_C of 53 MN $m^{-3/2}$ and a yield strength of 950 MN m^{-2}. Assuming that the plate contains cracks on the limit of detection, determine whether the plate will undergo general yield or will fail by fast fracture before general yielding occurs. What is the stress at which fast fracture would occur?

Answer: Failure by fast fracture at 946 MN m^{-2}.

30. Two wooden beams are butt-jointed using an epoxy adhesive (Fig. A1.3). The adhesive was stirred before application, entraining air bubbles which, under pressure in

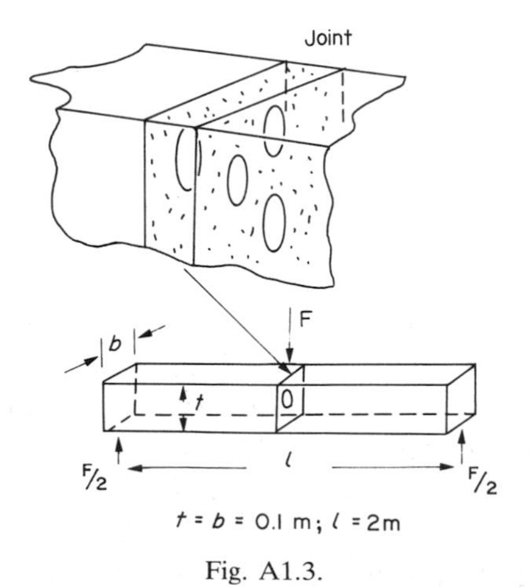

Fig. A1.3.

forming the joint, deform to flat, penny-shaped discs of diameter $2a = 2$ mm. If the beam has the dimensions shown, and epoxy has a fracture toughness of 0.5 MN $m^{-3/2}$, calculate the maximum load F that the beam can support. Assume $K = \sigma\sqrt{\pi a}$ for the disc-shaped bubbles.

Answer: 2.97 kN.

31. A component is made of a steel for which $K_C = 54\ \mathrm{MN\ m^{-3/2}}$. Non-destructive testing by ultrasonic methods shows that the component contains cracks of up to $2a = 0.2$ mm in length. Laboratory tests show that the crack-growth rate under cyclic loading is given by

$$\frac{\mathrm{d}a}{\mathrm{d}N} = A(\Delta K)^4,$$

where $A = 4 \times 10^{-13}\ (\mathrm{MN\ m^{-2}})^{-4}\ \mathrm{m^{-1}}$. The component is subjected to an alternating stress of range

$$\Delta\sigma = 180\ \mathrm{MN\ m^{-2}}$$

about a mean stress of $\Delta\sigma/2$. Given that $\Delta K = \Delta\sigma\sqrt{\pi a}$, calculate the number of cycles to failure.

Answer: 2.4×10^6 cycles.

32. When a fast-breeder reactor is shut down quickly, the temperature of the surface of a number of components drops from 600°C to 400°C in less than a second. These components are made of a stainless steel, and have a thick section, the bulk of which remains at the higher temperature for several seconds. The low-cycle fatigue life of the steel is described by

$$N_f^{1/2}\Delta\varepsilon^{\mathrm{pl}} = 0.2$$

where N_f is the number of cycles to failure and $\Delta\varepsilon^{\mathrm{pl}}$ is the plastic strain range. Estimate the number of fast shut-downs the reactor can sustain before serious cracking or failure will occur. (The thermal expansion coefficient of stainless steel is $1.2 \times 10^{-5}\ \mathrm{K^{-1}}$; the appropriate yield strain at 400°C is 0.4×10^{-3}.)

Answer: 10^4 shut-downs.

33. (a) An aluminium alloy for an airframe component was tested in the laboratory under an applied stress which varied sinusoidally with time about a mean stress of zero. The alloy failed under a stress range, $\Delta\sigma$, of $280\ \mathrm{MN\ m^{-2}}$ after 10^5 cycles; under a range of $200\ \mathrm{MN\ m^{-2}}$, the alloy failed after 10^7 cycles. Assuming that the fatigue behaviour of the alloy can be represented by

$$\Delta\sigma(N_f)^a = C$$

where a and C are material constants, find the number of cycles to failure, N_f, for a component subjected to a stress range of $150\ \mathrm{MN\ m^{-2}}$.

(b) An aircraft using the airframe components has encountered an estimated 4×10^8 cycles at a stress range of $150\ \mathrm{MN\ m^{-2}}$. It is desired to extend the airframe life by another 4×10^8 cycles by reducing the performance of the aircraft. Find the decrease in stress range necessary to achieve this additional life. You may assume a simple cumulative-damage law of the form

$$\sum_i \frac{N_i}{N_{fi}} = 1$$

for this calculation.

Indicate briefly how the following would affect the fatigue life of the components:
(a) a good surface finish;
(b) the presence of a rivet hole;
(c) a significant mean tensile stress;
(d) a corrosive atmosphere.

Answers: (a) 5.2×10^8 cycles, (b) 13 MN m^{-2}.

34. A cylindrical steel pressure vessel of 7.5 m diameter and 40 mm wall thickness is to operate at a working pressure of 5.1 MN m^{-2}. The design assumes that failure will take place by fast fracture from a crack which has extended gradually along the length of the vessel by fatigue. To prevent fast fracture, the total number of loading cycles from zero to full load and back to zero again must not exceed 3000.

The fracture toughness for the steel is 200 MN m$^{-3/2}$. The growth of the crack by fatigue may be represented approximately by the equation

$$\frac{\mathrm{d}a}{\mathrm{d}N} = A(\Delta K)^4,$$

where $A = 2.44 \times 10^{-14}$ (MN m^{-2})$^{-4}$ m^{-1}, $\mathrm{d}a/\mathrm{d}N$ is the extent of crack growth per load cycle and K is the stress intensity factor for the crack.

Find the minimum pressure to which the vessel must be tested before use to guarantee against failure in under the 3000 load cycles.

Answer: 8.97 MN m^{-2}.

35. Use your knowledge of diffusion to account for the following observations:
(a) Carbon diffuses fairly rapidly through iron at 100°C, whereas chromium does not.
(b) Diffusion is more rapid in polycrystalline silver with a small grain size than in coarse-grained silver.

Give an approximate expression for the time t required for significant diffusion to take place over a distance x in terms of x, and the diffusion coefficient, D. A component is made from an alloy of copper with 18% by weight of zinc. The concentration of zinc is found to vary significantly over distances of 10 μm. Estimate the time required for a substantial levelling out of the zinc concentration at 750°C. The diffusion coefficient for zinc in the alloy is given by

$$D = D_0 e^{-Q/\bar{R}T}$$

where $\bar{R}$ is the universal gas constant and T is the absolute temperature. The constants D_0 and Q have the values 9.5 mm^2 s^{-1} and 159 kJ mol^{-1}, respectively.

Answers: $t = x^2/D$; 23 min.

36. It is found that a force F will inject a given weight of a thermosetting polymer into an intricate mould in 30 s at 177°C and in 81.5 s at 157°C. If the viscosity of the polymer follows an *Arrhenius Law*, with a rate of process proportional to $e^{-Q/\bar{R}T}$, calculate how long the process will take at 227°C.

Answer: 3.5 s.

37. Explain what is meant by *creep* in materials. What are the characteristics of a creep-resistant material?

A cylindrical tube in a chemical plant is subjected to an excess internal pressure of 6 MN m^{-2}, which leads to a circumferential stress in the tube wall. The tube is required to withstand this stress at a temperature of 510°C for 9 years. A designer has specified tubes of 40 mm bore and 2 mm wall thickness made from a stainless alloy of iron with 15% by weight of chromium. The manufacturer's specification for this alloy gives the following information:

Temperature/°C	618	640	660	683	707
Steady-state creep rate $\dot{\varepsilon}$/s^{-1}, for an applied tensile stress σ of 200 MN m^{-2}	1.0×10^{-7}	1.7×10^{-7}	4.3×10^{-7}	7.7×10^{-7}	2.0×10^{-6}

Over the present ranges of stress and temperature the alloy can be considered to creep according to the equation

$$\dot{\varepsilon} = A\sigma^5 e^{-Q/\bar{R}T}$$

where A and Q are constants, $\bar{R}$ is the universal gas constant, and T is the absolute temperature. Given that failure is imminent at a creep strain of 0.01 for the present alloy, comment on the safety of the design.

Answer: Strain over 9 years = 0.00057; design safe.

38. Explain what is meant by *diffusion* in materials. Account for the variation of diffusion rates with (a) temperature, (b) concentration gradient and (c) grain size.

An alloy tie bar in a chemical plant has been designed to withstand a stress, σ, of 25 MN m^{-2} at 620°C. Creep tests carried out on specimens of the alloy under these conditions indicated a steady-state creep-rate, $\dot{\varepsilon}$, of 3.1×10^{-12} s^{-1}. In service it was found that, for 30% of the running time, the stress and temperature increased to 30 MN m^{-2} and 650°C. Calculate the average creep-rate under service conditions. It may be assumed that the alloy creeps according to the equation

$$\dot{\varepsilon} = A\sigma^5 e^{-Q/\surd\bar{R}T}$$

where A and Q are constants, $\bar{R}$ is the universal gas constant and T is the absolute temperature. Q has a value of 160 kJ mol^{-1}.

Answer: 6.82×10^{-12} s^{-1}.

39. The oxidation of a particular metal in air is limited by the outward diffusion of metallic ions through an unbroken surface film of one species of oxide. Assume that the concentration of metallic ions in the film immediately next to the metal is c_1, and that the concentration of ions in the film immediately next to the air is c_2, where c_1 and c_2 are constants. Use Fick's First Law to show that the oxidation of the metal should satisfy parabolic kinetics, with weight gain Δm given by

$$(\Delta m)^2 = k_P t.$$

The oxidation of another metal is limited by the outward flow of electrons through a uniform, unbroken oxide film. Assume that the electrical potential in the film immediately next to the metal is V_1, and the potential immediately next to the free surface is V_2, where V_1 and V_2 are constants. Use Ohm's Law to show that parabolic kinetics should apply in this case also.

40. The kinetics of oxidation of mild steel at high temperature are parabolic, with

$$k_P/\mathrm{kg^2\,m^{-4}\,s^{-1}} = 37 \exp\left\{-\frac{138\ \mathrm{kJ\,mol^{-1}}}{\bar{R}T}\right\}.$$

Find the depth of metal lost from the surface of a mild steel tie bar in a furnace at 500°C after 1 year. You may assume that the oxide scale is predominantly FeO. The atomic weight and density of iron are 55.9 kg kmol^{-1} and 7.87 Mg m^{-3}; the atomic weight of oxygen is 16 kg kmol^{-1}. What would be the loss at 600°C?

Answers: 0.33 mm at 500°C; 1.13 mm at 600°C.

41. Explain the following observations, using diagrams to illustrate your answer wherever you can.

(a) A reaction vessel for a chemical plant was fabricated by welding together stainless-steel plates (containing 18% chromium, 8% nickel and 0.1% carbon by weight). During service the vessel corroded badly at the grain boundaries near the welds.

(b) Mild-steel radiators in a central-heating system were found to have undergone little corrosion after several years' service.

(c) In order to prevent the corrosion of a mild-steel structure immersed in sea water, a newly qualified engineer suggested the attachment of titanium plates in the expectation of powerful cathodic action. He later found to his chagrin that the structure had corroded badly.

42. Explain the following observations, using diagrams to illustrate your answer wherever possible:

(a) Diffusion of aluminium into the surface of a nickel super-alloy turbine blade reduced the rate of high-temperature oxidation.

(b) Steel nails used to hold copper roofing sheet in position failed rapidly by wet corrosion.

(c) The corrosion of an underground steel pipeline was greatly reduced when the pipeline was connected to a buried bar of magnesium alloy.

43. (a) Measurements of the rate of crack growth in brass exposed to ammonium sulphate solution and subjected to a constant tensile stress gave the following data:

Nominal stress σ/MN m^{-2}	Crack depth a/mm	Crack growth rate da/dt/mm year^{-1}
4	0.25	0.3
4	0.50	0.6
8	0.25	1.2

Show that these data are consistent with a relationship of form

$$\frac{da}{dt} = AK^n$$

where $K = \sigma\sqrt{\pi a}$ is the stress intensity factor. Find the values of the integer n and the constant A.

(b) The critical strain energy release rate, G_C, for brass in the present environment is 55 kJ m^{-2}, with a Young's modulus of 110 GN m^{-2}. It is proposed to use the brass for piping in an ammonium sulphate plant. The pipes must sustain a circumferential tensile stress of 85 MN m^{-2}, and experience has shown that longitudinal scratches 0.02 mm deep are likely to occur on the inner surfaces of the pipes. Estimate the time that a pipe would last without fracturing once the solution started to flow through it.

(c) How might you protect the inside of the pipe against chemical attack?

Answers: (a) $n = 2$, $A = 0.0239$ m^4 MN^{-2} year^{-1}; (b) 6.4 days.

44. Under aggressive corrosion conditions it is estimated that the maximum corrosion current density in a galvanised steel sheet will be 6×10^{-3} A m^{-2}. Estimate the thickness of the galvanised layer needed to give a rust-free life of at least 5 years. The density of zinc is 7.13 Mg m^{-3}.

Answer: 0.045 mm.

45. A sheet of steel of thickness 0.50 mm is tinplated on both sides and subjected to a corrosive environment. During service, the tinplate becomes scratched, so that steel is exposed over 0.5% of the area of the sheet. Under these conditions it is estimated that the current consumed at the tinned surface by the oxygen-reduction reaction is 2×10^{-3} A m^{-2}. Will the sheet rust through within 5 years in the scratched condition? The density of steel is 7.87 Mg m^{-3}.

Answer: Yes.

46. (a) Explain the origins of friction between solid surfaces in contact.

(b) Soft rubbers do not obey the law of friction $F_s = \mu_s P$ (where F_s is the sliding force, P the normal force acting across the surfaces and μ_s the coefficient of static friction). Instead, F_s increases with the nominal contact area A (for this reason racing cars have wide tyres). Explain this.

(c) How does lubrication reduce friction? How can friction between road and tyre be maintained even under conditions of appreciable lubrication?

47. It is observed that snow lies stably on roofs with a slope of less than 24°, but that it slides off roofs with a greater slope. Skiers, on the other hand, slide on a snow-covered mountain side with a slope of only 2°. Why is this?

A man of weight 100 kg standing on skis 2 m long and 0.10 m wide slides on the 2° mountain slope, at 0°C. Calculate the work done against friction when the ski slides a distance equal to its own length. Hence calculate the average thickness of the water film beneath each ski. (The latent heat of fusion of ice is 330 MJ m^{-3}.)

Answers: Work done 69 J; average film thickness = 0.5 μm.

APPENDIX 2

AIDS AND DEMONSTRATIONS

The following is a summary of visual aids (slides, artefacts and demonstrations) that may be found helpful in presenting the material in this book. Material for slides may be found in this book; in the further reading at the ends of the chapters; and in other readily available sources (indicated by references [1] to [5] and listed at end of Appendix 2). Where material for slides needs to be found from more specialised publications and reports we have given appropriate references. Copyright permission should, of course, be obtained where applicable.

Chapter 1

Slides: S.S. *Schenectady* after fast fracture in dock [1]; sectioned drawing of turbofan aero-engine [2]; sectioned drawing of spark plug; sailing cruisers (from yachting magazines); bridges.

Artefacts: Screwdrivers; dismantled spark plug; PVC yachting anorak, polymer rope, etc.

Chapter 2

Slides: Map of World (to illustrate strategic factors); open-cast copper mine (to emphasise energy needed to extract ores); recycling of scrap metals, glass, building materials, etc.

Chapter 3

Slides: Vaulting pole, springs, girders (for extremes of stiffness).

Demonstrations: (a) Foam polyurethane $\approx 40 \times 5 \times 5$ cm pulled along its length in tension. (b) Foam $\approx 60 \times 60 \times 10$ cm glued into square wooden frame hinged at all four corners and sheared. (c) Foam $\approx$20-cm cube loaded in compression by 10-kg weight on wooden platten to give $\approx$4% strain; $E \approx 10^{-4}$ GN m^{-2}. (d) Rods of steel, glass, wood $\approx$ 6 mm diameter $\times$ 0.75 m long suspended by string at each end (Fig. 3.4) with 0.5-kg weight at mid point; f values $\approx$ 10, 6 and 2 s^{-1}.

Chapter 4

Demonstrations: (a) Atom spring models (Fig. 4.2) on overhead projector to illustrate effect of structure on modulus. (b) Large models of Na atom and Cl atom. (c) Liquid nitrogen.

Chapter 5

Demonstrations: (a) Give four injection-moulded close-packed planes to each student to allow personal building of f.c.c. and c.p.h. (b) Atomix atomic model on overhead projector to show atom packing (Emotion Productions Inc., 4825 Sainte Catherine O, Montreal 215PQ, Canada); or ball bearings on overhead projector.

Chapter 6

Slides: Microstructures of GFRP, glass-filled polymer, cermet, wood; sectioned piece of cord-reinforced automobile tyre.

Demonstrations: (a) Put 15-mm rubber tube into vacuum flask of liquid nitrogen—tube should have steel rod inside to keep it straight. Take tube out after 3 min and remove steel rod (wearing gloves). Support tube horizontally with lagged support at each end. Load centre of span with 0.5-kg weight and lagged hook. Rubber will become floppy after 1–2 min. (b) Glue alternate sheets of foam polyurethane and plywood together to make a sandwich composite—stiff one way, very floppy the other.

Chapter 7

Slides: Of reflecting telescopes, aeroplanes, space capsules, bicycles (to illustrate applications of stiff but light materials).

Chapter 8

Slides: Slab and sheet metal rolling; extrusion, etc., of polymers; tensile-testing machines; hardness-testing machines; hardness indentations.

Demonstrations: (a) Pull rubber band on overhead projector to show large elastic strain. (b) Take a piece of plasticine modelling material (Harbutt Ltd., Bathampton, Bath BA2 6TA, England; from most toy shops) and roll into a rod ≃ 2.5 cm diameter × 12 cm long. Form central portion to give slightly reduced gauge section. Pull on overhead projector to show elastic and plastic deformation and necking.

Chapter 9

Demonstrations: (a) Take offcut of carpet ≈0.5 × 3 m; put on bench and pass rucks along (Fig. 9.6). (b) Raft of pencils on overhead projector to simulate plank analogy (Fig. 9.10).

Chapter 10

Slides: Microstructures showing precipitates; electron micrographs of dislocation tangles; micrographs of polycrystalline metals.

Demonstrations: (a) Atomix (to show grain boundaries). (b) Model of dispersion strengthening. Take piece of PMMA sheet ≃ 2.5 mm thick and ≃ 7 cm square. Glue four PMMA strips of section ≃ 7 × 7 mm on top of the sheet to form a tray ≃ 7 mm deep. Cut. six ≃ 7-mm lengths of an ≃ 6-mm-diameter PMMA rod. Glue the ends of these to the bottom of the tray to form a line of "stepping stones" spaced equally (≃3 mm) apart and going straight across the centre of the tray from edge to edge. Cap the top of each stepping stone with a 6-mm self-adhesive disc. Lightly grease the inside of the tray and also the stepping stones. Place tray on overhead projector. Gently pour coloured water into one side of tray until water is near stepping-stone obstacles. Tilt tray up slightly to allow water to run up to obstacles. Show bowing of water meniscus between obstacles, with eventual breakthrough; the surface tension of the water is analogous to the line tension of a dislocation, and regions where the water has broken through are analogous to regions of the crystal that have undergone plastic deformation by amount b.

Chapter 11

Slides: Necked tensile specimens of metals; deep drawing operations and deep-drawn cans, etc.

Demonstrations: (a) Push two very blunt wooden wedges together on overhead projector (Fig. 11.1) to show shearing under compressive loading. (b) Make two-dimensional working model of Fig. 11.4 out of PMMA sheet for use on overhead projector (note—remove offending apexes to allow movement, and put markers on either side of shear planes to show up the shear displacements on the overhead). (c) Necking of plasticine on overhead projector (see demonstration in Chapter 8). (d) Stable necking of polyethylene. Cut a gauge length ≃ 7 × 70 mm from polyethylene sheet (length of specimen parallel to roll direction of sheet). Pull in tension on overhead, and observe propagation of stable neck.

Chapter 12

Slides: Springs of various types; multi-leaf springs on trucks, automobiles, steam locomotives, etc.; light pressure vessels—e.g. aeroplane fuselages; cheap pressure vessels—e.g. water tanks, nuclear reactor vessels; metal rolling stand.

Demonstrations: Take a strip ≃ 0.25 mm × 1 cm × 15 cm of cold-rolled (work-hardened) brass and bend it (on edge) on the overhead until permanent deformation takes place. Anneal brass strip at bright red heat for ≃0.5 min to soften it.* After cooling replace on overhead and show that permanent deformation takes place at a much smaller deflection than before. This illustrates the importance of large σ_y in springs.

*For this and subsequent demonstrations involving a heat source, use a gas torch such as the Sievert self-blowing propane outfit (W. A. Meyer Ltd., 9/11 Gleneldon Road, London SW16 2AU, or from most tool shops): this comprises 3.9-kg propane bottle, 3085 hose-failure valve, fitted pressure hose no. 16310, 3486 torch, 2941 burner.

Chapter 13

Slides: Fast-structure failures in ships [1], pressure vessels, pipelines, flywheels, etc.

Demonstrations: (a) Balloons and safety pin (see Chapter 13, p. 121). Afterwards, put fractured edges of balloon rubber on overhead to show that wavy fracture path closely parallels that seen when metals have undergone fast fracture. (b) G_c for Sellotape (see Chapter 13, p. 122).

Chapter 14

Slides: Plastic cavitation around inclusions in metals (e.g. metallographic section through neck in tensile specimen); SEM pictures of fracture surfaces in ductile metals, glass, alkali halide crystals.

Artefacts: Damaged piece of GFRP to show opacity caused by debonding.

Demonstrations: (a) Take piece of plasticine $\simeq 1 \times 3 \times 10$ cm. Using knife put notch $\simeq 5$ mm deep into long edge. Pull on overhead and watch notch tip blunting by plastic flow. (b) Pull plasticine to failure to show high toughness and rough fracture surface. (c) Notch $\simeq$5-mm-diameter glass rod with sharp triangular file and break on overhead to show low toughness and smooth fracture surface. (d) Put rubber tube in liquid nitrogen for $\simeq 2$ min; remove and smash with hammer *behind safety screens* to show low toughness. (e) Heat an $\simeq$2-mm-diameter medium carbon steel rod to bright red and quench into water. Using fingers, snap rod on overhead to show low toughness. Harden a second rod, but reheat it to give a light blue colour. Show that this tempering makes it much harder to snap the rod (use thick gloves and safety glasses in (c), (d) and (e) and put a safety screen between these demonstrations and the audience).

Chapter 15

Slides: Fatigue fracture surfaces; components failed by fatigue, e.g. gear teeth, half-shafts, etc.

Demonstration: Make a pendulum comprising some 5-A fuse wire hanging from a horizontal knife-edge and carrying a 0.5-kg weight. Make the length of the fuse wire in pendulum $\simeq 15$ cm. Make the weight oscillate with amplitude $\simeq 7$ cm. Flexing of the wire where it passes over the knife edge will lead to fatigue failure after $\simeq 1$ min.

Chapter 16

Slides: Ultrasonic crack detection; hydraulic testing.

Chapter 17

Slides: Tungsten filaments, turbine blades, lead drain pipes and organ pipes, glaciers; creep-testing rigs; micrographs of creep cavities.

Demonstrations: (a) Wind $\simeq$2-cm-diameter, 8-cm-long coil of $\simeq$1.5-mm-diameter Pb–Sn solder. Suspend coil from one end and observe marked creep extension of coil after $\simeq 15$ min at room temperature. (b) Observe self-weight creep of $\simeq$45-cm length of $\simeq$1-cm-diameter polyethylene tube held horizontally at one end. (c) Support an $\simeq$2-mm-diameter steel wire horizontally at one end. Hang a 20-g weight from the free end. Support a second identical length of wire immediately alongside the first, and hang a 40-g weight from its free end. Heat the pair of wires to red heat at their clamped ends and observe creep; note that the creep rate of the second wire is much more than twice that of the first, illustrating power-law creep.

Chapter 18

Demonstrations: (a) Inject a drop of coloured dye under the surface of a very shallow stagnant pool of water in a Petri dish on the overhead. Observe dye spreading by diffusion with time. (b) Atomix to show vacancies, and surface diffusion.

Chapter 19

Demonstration: Fit up a dashpot and spring model (Fig. 19.7) and hang it from a support. Hang a weight on the lower end of the combination and, using a ruler to measure extension, plot the creep out on the blackboard. Remove weight and plot out the reverse creep.

Chapter 20

Slides: Turbofan aero-engine; super-alloy turbine blades, showing cooling ports [3]; super-alloy microstructures [4]; DS eutectic microstructures [3, 5]; ceramic turbine blades.

Chapter 21

Slides: Pitting corrosion on a marine turbine blade [4]; corroded tie bars, etc., in furnaces, heat exchangers, etc.; oxidised cermets.

Demonstrations: (a) Heat one end of a $\simeq 2 \times 5 \times 120$-mm piece of lightly abraded mild steel in the gas flame. After $\simeq 1$ min at bright red heat, plunge into cold water in a dish on the overhead. Oxide flakes will spall off into the water leaving a bright metal surface. (b) Lightly abrade a $\simeq$0.1-mm $\times$ 5-cm $\times$ 5-cm piece of copper shim. Play the gas flame on one side of it, using the reducing region of the flame, and keep at medium red heat for $\simeq 1$ min. Then plunge immediately into cold water. Place shim on edge on overhead to show pronounced bending effect. This shows the effect of oxygen partial pressure on oxidation rate. The metal in contact with the reducing flame has a negligibly thin oxide layer; the oxide layer on the other side, where oxygen was available, is quite thick. The differential thermal contraction between this thick layer and the copper has caused a "bimetallic strip" effect.

Chapter 22

Slides: Microstructures of oxide layers and oxide-resistant coatings on metals and alloys; selective attack of eutectic alloys [5].

Demonstrations: Take a piece of 0.1-mm × 5-cm × 5-cm stainless-steel shim, and a similar piece of mild-steel shim. Degrease, and weigh both. Heat each for ≈1 min in the gas flame to bright red heat. The mild-steel shim will gain weight by more than ≈0.05 g. The stainless-steel shim will not gain weight significantly.

Chapter 23

Slides: Corroded automobiles, fences, roofs; stress-corrosion cracks, corrosion-fatigue cracks, pitting corrosion.

Demonstrations: Mix up an indicator solution as follows: dissolve 5 g of potassium ferricyanide in 500 cm^3 distilled water. Dissolve 1 g of phenolphthalein in 100 cm^3 ethyl alcohol. Take 500 cm^3 of distilled, aerated water and to it add 5 g sodium chloride. Shake until dissolved. Add 15 cm^3 of the ferricyanide solution and shake. Gradually add 45 cm^3 of phenolphthalein solution, shaking all the time (but stop adding this if the main solution starts to go cloudy). (a) Pour indicator solution into a Petri dish on the overhead. Degrease and lightly abrade a steel nail and put into dish. After ≈10 min a blue deposit will form by the nail, produced by reaction between Fe^{++} and the ferricyanide and showing that the iron is corroding. A pink colour will also appear, produced by reaction between OH^- and phenolphthalein, and showing that the oxygen-reduction reaction is taking place. (b) Modify a voltmeter so that the needle can be seen when put on the overhead. Wire up to galvanic couples of metals such as Cu, Fe, Zn, and Pt foil in salty water and show the voltage differences.

Chapter 24

Slides: Covering pipelines with polymeric films; cathodic protection of pipelines, ships, etc., with zinc bracelets; use of inert polymers in chemical plant; galvanic corrosion in architecture (e.g. Al window frames held with Cu bolts); weld decay.

Artefacts: Galvanised steel sheet, new and old; anodised Al; polymeric roofing material; corroded exhaust system.

Demonstrations: (a) Put indicator solution in Petri dish on overhead. Take steel nail and solder a Zn strip to it. Degrease, lightly abrade, and put in solution. No blue will appear, showing that the Fe is cathodically protected by the Zn. Pink will appear due to OH^- (produced by the oxygen-reduction reaction) because the Zn is corroding. (b) Put two degreased and lightly abraded steel nails in indicator solution on overhead. Wire a 4.5-V battery across them. Observe blue at one nail, pink at the other. This illustrates imposed-potential protection. (c) Solder a piece of Cu to a steel nail. Degrease, lightly abrade, and put in solution on overhead. Observe rapid build-up of blue at nail, pink at Cu, showing fast corrosion produced by mixing materials having different wet corrosion voltages.

Chapter 25

Slides: Bearings; brake linings; grinding and metal-cutting operations; taper sections of metal surfaces.

Demonstrations: (a) Block on inclined plane to determine μ. (b) Make a pair of rough surfaces from plasticine. Press together on overhead to show junction deformation. Shear on overhead to show origin of frictional force. (c) Gouge fragments out of plasticine on overhead using a serrated piece of wood to simulate abrasive wear. (d) Write on blackboard using chalk. Light "pressure" leaves little chalk on board—heavy "pressure" leaves a lot of chalk, showing dependence of adhesive wear rate on contact force.

Chapter 26

Slides: Split-shell bearings; hard particles embedded in soft bearing alloys; micrograph of section through layered bearing shell; skiers; automobile tyres.

Artefacts: Skis sectioned to show layered construction.

Demonstrations: (a) Put lump of plasticine between plattens of hand-operated hydraulic press. Monitor compressive straining of plasticine with a dial gauge. Plot load against compression on the blackboard. Show how plastic constraint when the plasticine is squashed down to a very thin layer vastly increases the load it can support. (b) Take a piece ≈1.5 × 5 × 5-cm low-loss rubber. Show that it is low loss by dropping an ≈3-cm-diameter steel ball on to it, giving large rebound. Repeat with a piece of high-loss rubber, giving little rebound. Make an inclined plane of frosted glass, and soap it. Place the pads of rubber at the top of the plane, and adjust angle of plane until low-loss pad slides rapidly downhill but high-loss pad slides only slowly if at all. (It is worth spending some extra time in building a pair of toy clockwork-driven tractors, one shod with low-loss tyres, the other with high-loss. Provided the slope of the ramp is suitably adjusted, the low-loss tractor will be unable to climb the soaped slope, but the high-loss one will.)

Chapter 27

Slides: Cars; steel-pressing plant; car assembly line; hand lay-up of GFRP; polymer moulding plant.

1. *Final Report of a Board of Investigation—The Design and Methods of Construction of Welded Steel Merchant Vessels*, Government Printing Office, Washington, DC, USA, 1947.
2. Coloured wall chart obtainable from Rolls-Royce Ltd., P.O. Box 31, Derby DE2 8BJ, England.
3. *Current and Future Materials Usage in Aircraft Gas Turbine Engines*, Metals and Ceramics Information Center, Battelle Laboratories, 505 King Avenue, Columbus, Ohio 43201, USA.
4. *The Nimonic Alloys*, 2nd edition, W. Betteridge and J. Heslop, Arnold, 1974.
5. *Conference on In-Situ Composites* II, edited by M. R. Jackson, J. L. Walter, F. D. Lemkey and R. W. Hertzberg, Xerox Publishing, 191 Spring Street, Lexington, Mass. 02173, USA, 1976.

APPENDIX 3

SYMBOLS AND FORMULAE

List of principal symbols

Symbol	*Meaning (units)*	*Where defined or first used*
Note:	Multiples or sub-multiples of basic units indicate the unit suffixes typically used with materials data.	
a	side of cubic unit cell (nm)	Chapter 5, Fig. 5.3
a	crack length (mm)	Chapter 13, p. 122
a	constant in Basquin's Law (dimensionless)	Chapter 15, p. 136
A	constant in fatigue crack-growth law	Chapter 15, p. 139
A	constant in creep law $\dot{\varepsilon}_{ss} = A\sigma^n e^{-Q/\bar{R}T}$	Chapter 17, p. 161
$\mathbf{b}$	Burgers vector (nm)	Chapter 9, Fig. 9.4
b	constant in Coffin–Manson Law (dimensionless)	Chapter 15, p. 137
c	concentration (m^{-3})	Chapter 18, p. 166
C_1	constant in Basquin's Law ($MN\,m^{-2}$)	Chapter 15, p. 136
C_2	constant in Coffin–Manson Law (dimensionless)	Chapter 15, p. 137
D	diffusion coefficient ($m^2\,s^{-1}$)	Chapter 18, p. 166
D_0	pre-exponential constant in diffusion coefficient ($m^2\,s^{-1}$)	Chapter 18, p. 169
E	Young's modulus of elasticity ($GN\,m^{-2}$)	Chapter 3, p. 29
f	force acting on unit length of dislocation line ($N\,m^{-1}$)	Chapter 9, p. 94
F	force (N)	Chapter 3, Fig. 3.1
g	acceleration due to gravity on the Earth's surface ($m^2\,s^{-1}$)	Chapter 7, p. 64
G	shear modulus ($GN\,m^{-2}$)	Chapter 3, p. 29
G_c	toughness (or critical strain energy release rate) ($kJ\,m^{-2}$)	Chapter 13, p. 122
H	hardness ($kg\,mm^{-2}$)	Chapter 8, Fig. 8.13
J	diffusion flux ($m^{-2}\,s^{-1}$)	Chapter 18, p. 166
k	shear yield strength ($MN\,m^{-2}$)	Chapter 11, p. 105
k	Boltzmann's constant, $\bar{R}/N_A$ ($J\,K^{-1}$)	Chapter 18, p. 167
K	bulk modulus ($GN\,m^{-2}$)	Chapter 3, p. 29
K	stress intensity factor ($MN\,m^{-3/2}$)	Chapter 13, p. 125
K_c	fracture toughness (critical stress intensity factor) ($MN\,m^{-3/2}$)	Chapter 13, p. 125

ΔK	K range in fatigue cycle ($MN\,m^{-3/2}$)	Chapter 15, p. 139
m	constant in fatigue Crack Growth Law (dimensionless)	Chapter 15, p. 139
n	creep exponent in $\dot{\varepsilon}_{ss} = A\sigma^n e^{-Q/\bar{R}T}$	Chapter 17, p. 160
N	number of fatigue cycles	Chapter 15, p. 136
N_A	Avogadro's number (mol^{-1})	Chapter 18, p. 169
N_f	number of fatigue cycles leading to failure (dimensionless)	Chapter 15, p. 136
$\tilde{p}$	price of material (UK£ or US\$ $tonne^{-1}$)	Chapter 2, Table 2.1
Q	activation energy per mole ($kJ\,mol^{-1}$)	Chapter 17, p. 161
r_0	equilibrium interatomic distance (nm)	Chapter 4, Fig. 4.4
$\bar{R}$	universal gas constant ($J\,K^{-1}\,mol^{-1}$)	Chapter 17, p. 161
S_0	bond stiffness ($N\,m^{-1}$)	Chapter 4, p. 41
t_f	time-to-failure (s)	Chapter 17, p. 163
T	line tension of dislocation (N)	Chapter 9, p. 95
T	absolute temperature (K)	Chapter 17, p. 158
T_M	absolute melting temperature (K)	Chapter 17, p. 158
U^{el}	elastic strain energy (J)	Chapter 8, Fig. 8.1
γ	(true) engineering shear strain (dimensionless)	Chapter 3, Fig. 3.3
Δ	dilatation (dimensionless)	Chapter 3, Fig. 3.3
ε	true (logarithmic) strain (dimensionless)	Chapter 8, p. 76
ε_f	(nominal) strain after fracture; tensile ductility (dimensionless)	Chapter 8, Table 8.1
ε_n	nominal (linear) strain (dimensionless)	Chapter 3, Fig. 3.3
ε_0	permittivity of free space ($F\,m^{-1}$)	Chapter 4, Fig. 4.3
$\dot{\varepsilon}_{ss}$	steady-state tensile strain-rate in creep (s^{-1})	Chapter 17, p. 160
$\Delta\varepsilon^{pl}$	plastic strain range in fatigue (dimensionless)	Chapter 15, Fig. 15.3
μ_k	coefficient of kinetic friction (dimensionless)	Chapter 25, p. 224
μ_s	coefficient of static friction (dimensionless)	Chapter 25, p. 223
ν	Poisson's ratio (dimensionless)	Chapter 3, Fig. 3.3
ρ	density ($Mg\,m^{-3}$)	Chapter 5, Table 5.1
σ	true stress ($MN\,m^{-2}$)	Chapter 3, Fig. 3.1
σ_n	nominal stress ($MN\,m^{-2}$)	Chapter 8, p. 77
σ_{TS}	(nominal) tensile strength ($MN\,m^{-2}$)	Chapter 8, Fig. 8.11
σ_y	(nominal) yield strength ($MN\,m^{-2}$)	Chapter 8, Fig. 8.11
$\tilde{\sigma}$	ideal strength ($GN\,m^{-2}$)	Chapter 9, p. 87
$\Delta\sigma$	stress range in fatigue ($MN\,m^{-2}$)	Chapter 15, p. 136
τ	shear stress ($MN\,m^{-2}$)	Chapter 3, Fig. 3.1

Summary of principal formulae and of magnitudes

Chapter 2: *Exponential Growth*

$$\frac{dC}{dt} = \frac{rC}{100}$$

C = consumption rate (tonne $year^{-1}$); r = fractional growth rate (% $year^{-1}$); t = time.

Chapter 3: *Definition of Stress, Strain, Poisson's Ratio, Elastic Moduli*

$$\sigma = \frac{F}{A} \qquad \tau = \frac{F_s}{A} \qquad p = -\frac{F}{A} \qquad \nu = -\frac{\text{lateral strain}}{\text{tensile strain}}$$

$$\varepsilon_n = \frac{u}{l} \qquad \gamma = \frac{w}{l} \qquad \Delta = \frac{\Delta V}{V}$$

$$\sigma = E\varepsilon_n \qquad \tau = G\gamma \qquad p = -K\Delta$$

$F(F_s)$ = normal (shear) component of force; A = area; $u(w)$ = normal (shear) component of displacement; $\sigma(\varepsilon_n)$ = true tensile stress (nominal tensile strain); $\tau(\gamma)$ = true shear stress (true engineering shear strain); $p(\Delta)$ = external pressure (dilatation); ν = Poisson's ratio; E = Young's modulus; G = shear modulus; K = bulk modulus.

Chapter 8: *Nominal and True Stress and Strain, Energy of Deformation*

$$\sigma_n = \frac{F}{A_0}; \qquad \sigma = \frac{F}{A}; \qquad \varepsilon_n = \frac{u}{l_0} = \frac{l - l_0}{l_0}; \qquad \varepsilon = \int_{l_0}^{l} \frac{dl}{l} = \ln\left(\frac{l}{l_0}\right),$$

$A_0 l_0 = Al$ for *plastic* deformation; or for elastic or elastic/plastic deformation when $\nu = 0.5$. Hence

$$\sigma = \sigma_n (1 + \varepsilon_n).$$

Also

$$\varepsilon = \ln(1 + \varepsilon_n).$$

Work of deformation, per unit volume:

$$U = \int_{\varepsilon_{n1}}^{\varepsilon_{n2}} \sigma_n \, d\varepsilon_n = \int_{\varepsilon_1}^{\varepsilon_2} \sigma \, d\varepsilon.$$

For linear-elastic deformation *only*

$$U = \frac{\sigma_n^2}{2E}.$$

Hardness,

$$H = F/A,$$

σ_n = nominal stress, $A_0(l_0)$ = initial area (length), $A(l)$ = current area (length); ε = true strain.

Chapters 9 and 10: *Dislocations*

The dislocation yield-strength,

$$\tau_y = \frac{\tilde{c}T}{\mathbf{b}L}$$

$$\sigma_y = 3\tau_y,$$

T = line tension (about $Gb^2/2$); $\mathbf{b}$ = Burgers vector; L = obstacle spacing; $\tilde{c}$ = constant ($\tilde{c} = 2$ for strong obstacles; $\tilde{c} < 2$ for weak obstacles); σ_y = yield strength.

Chapter 11: *Plasticity*

Shear yield stress,

$$k = \sigma_y/2.$$

Hardness,

$$H \approx 3\sigma_y.$$

Necking starts when

$$\frac{d\sigma}{d\varepsilon} = \sigma.$$

Chapters 13 and 14: *Fast Fracture*

The stress intensity

$$K = Y\sigma\sqrt{\pi a}; \; Y \approx 1.$$

Fast fracture occurs when

$$K = K_c = \sqrt{EG_c},$$

a = crack length; Y = dimensionless constant; K_c = critical stress intensity or fracture toughness; G = critical strain energy release rate or toughness.

Chapter 15: *Fatigue*

No pre-cracks

Basquin's Law (high cycle)

$$\Delta\sigma N_f^a = C_1.$$

Coffin–Manson Law (low cycle)

$$\Delta\varepsilon^{pl} N_f^b = C_2.$$

Goodman's Rule

$$\Delta\sigma \text{ (for } \sigma_m = \sigma_m) = \Delta\sigma \text{ (for } \sigma_m = 0) \left\{1 - \frac{|\sigma_m|}{\sigma_{TS}}\right\}.$$

Miner's Rule for cumulative damage

$$\sum_i \frac{N_i}{N_{fi}} = 1.$$

For pre-cracked materials

Crack Growth Law

$$\frac{da}{dN} = A\Delta K^m.$$

Failure by crack growth

$$N_f = \int_{a_0}^{a_f} \frac{da}{A\Delta K^m},$$

$\Delta\sigma$ = stress range; $\Delta\varepsilon^{pl}$ = plastic strain range; ΔK = stress intensity range; N = cycles; N_f = cycles to failure; C_1, C_2, a, b, A, m = constants; σ_m = mean stress; σ_{TS} = tensile strength; a = crack length.

Chapter 17: *Creep and Creep Fracture*

$$\dot{\varepsilon}_{ss} = A\sigma^n e^{-Q/\bar{R}T},$$

$\dot{\varepsilon}_{ss}$ = steady-state tensile strain-rate; Q = activation energy; $\bar{R}$ = universal gas constant; T = absolute temperature; A, n = constants.

Chapter 18: *Kinetic Theory of Diffusion*

Fick's Law

$$J = -D\frac{dc}{dx}.$$

Arrhenius's Law

$$\text{Rate} \propto e^{-Q/\bar{R}T}.$$

Diffusion coefficient

$$D = D_0 e^{-Q/\bar{R}T},$$

J = diffusive flux; D = diffusion coefficient; c = concentration; x = distance; D_0 = pre-exponential factor.

Chapter 21: *Oxidation*

Linear Growth Law

$$\Delta m = k_L t; \qquad k_L = A_L e^{-Q/\bar{R}T}.$$

Parabolic Growth Law

$$\Delta m^2 = k_P t; \qquad k_P = A_P e^{-Q/\bar{R}T}.$$

Δm = mass gain per unit area; k_L, k_P, A_L, A_P = constants.

Chapter 25: *Friction and Wear*

True contact area

$$a \approx P/\sigma_y.$$

P = contact force.

Magnitudes of properties

The listed properties lie, for most structural materials, in the range shown.

Moduli of elasticity, E	2 to 200 $GN\,m^{-2}$
Densities, ρ	1 to 10 $Mg\,m^{-3}$
Yield strengths, σ_y	20 to 200 $MN\,m^{-2}$
Toughnesses, G_c	0.2 to 200 $kJ\,m^{-2}$
Fracture toughnesses, K_c	0.2 to 200 $MN\,m^{-3/2}$

INDEX